CROSS-PLATFORM WEB
SERVICES USING
C# AND JAVA

CROSS-PLATFORM WEB SERVICES USING C# AND JAVA

BRIAN HOCHGURTEL

CHARLES RIVER MEDIA, INC.
Hingham, Massachusetts

Acquisitions Editor: Jim Walsh
Production: Publishers' Design and Production Services, Inc.
Cover Design: The Printed Image

CHARLES RIVER MEDIA, INC.
10 Downer Avenue
Hingham, Massachusetts 02043
781-740-0400
781-740-8816 (FAX)
info@charlesriver.com
www.charlesriver.com

This book is printed on acid-free paper.

Brain Hochgurtel. *Cross-Platform Web Services Using C# and Java*.
ISBN: 1-58450-262-2

Library of Congress Cataloging-in-Publication Data

Hochgurtel, Brian.
   Cross-platform Web services using C# and Java / Brian Hochgurtel.
      p.   cm.
   ISBN 1-58450-262-2
   1. Web site development.   2. Internet programming.   3. C (Computer
program language)   4. Java (Computer program language)   5. Cross-platform
software development.   I. Title.
   TK5105.888.H617 2003
   005.2'76—dc21
                              2002154754
Printed in the United States of America
03 7 6 5 4 3 2   First Edition

Contents

Acknowledgments

First of all, I need to thank my wife for putting up with me hiding in my office for six months while I wrote this book. Her patience and support were astounding.

I also need to thank my friend, Rod Stephens, for taking me under his wing during the book we wrote together in 2001. Without him I doubt I could have written or even proposed this book.

Preface

The goal of this book is to provide you with an introduction to Web Services under C# and Java, along with the underlying standards that drive them. This book shows how the different technologies work and collaborate, and then describes security and examines more practical examples.

STRUCTURE OF THE BOOK

In the first section, the book examines the underlying XML standards and the concepts of remote objects. Many times the XML that works underneath Web Services can be ignored because the software operates at such a high level. But when you run into glitches or you just happen to look at a request and response within a viewer, it's useful to know what the XML represents and the function it serves.

In the second section, Java and C# technologies that deploy Web Services are examined and several examples are given. This involves not only creating the service but also creating software and Web pages that act as a consumer.

In the third section, the details of deploying Web Services are examined. This includes using tools to make Web Services from C# communicate with Java and visa versa. This section also looks closely at gathering and implementing the security requirements necessary to protect Web Services.

Most of the Web Service examples in the book are quick and to the point, but the third section the book takes a closer look at the deployment and testing of a more complex Web Service. This allows you to examine how the Web Service works with the various consumers and how to pass more complex values.

HOW TO USE THIS BOOK

This book builds on itself as you move forward. This first section covers the basics of the underlying standards that make Web Services work. If you have a basic or general understanding of these standards, you may wish to move on to the second and third sections and refer back to these introductory chapters as needed.

Many of the examples in the text are fairly self-contained so that if you need a quick reference on how to create a particular Web Service consumer or service you can find it quickly. As the book examines the different technologies, each chapter contains examples that are similar so there is an easy cross-reference.

The book assumes that you have some understanding of how to compile Java and C# programs. The examples in each chapter do explain where to put certain files and how to set environmental settings, but you may want to have introductory texts lying around so you have a quick reference to items such as the *Java Virtual Machine* (JVM) or the *Common Language Runtime* (CLR).

WHAT YOU NEED

This book requires that you have a PC running with either *NT, Windows 2000,* or *Windows XP Professional* or *Server* version so you can utilize Microsoft's Internet Information Server (IIS). Note that the professional versions of these operating systems will not allow you to use SSL, but you can use Apache with SSL on any of them.

You will also need either *Visual Studio.NET* or the *.NET Framework Software Development Kit* to compile and execute the C# examples.

To compile and run the Java examples, you need a *Java Development Kit* for compilation and a Servlet container to execute the examples. The book uses *Tomcat* from the Apache group as the container but any other container such as *Sun One Server* or BEA's *Weblogic* should work just fine. Note that you don't need the professional versions of the *Windows* operating systems to use the Java examples, but without them you'll be limited to just Java and won't be able to use any of the cross-platform examples.

Appendix B contains the URLs for downloading the various software needed for this book.

ASSUMED KNOWLEDGE

This book assumes that you have experience dealing with objects and object-oriented programming. The examples in the book are straightforward, but the assumption is that you understand concepts such as a constructor, instantiating an object, and methods.

Introduction to Web Service Technologies

XML and the concept of objects and remote objects are the driving force behind Web Services. The concept of Web Services comes from the idea of *Remote Procedure Call* (RPC), which is a simple way of calling methods from across a network. The problem with RPC was that every vendor had a different standard for transporting the data. By using XML, the data in Web Services now moves in a predictable manner across a network or the Web.

Section I focuses on introducing you to the concepts and standards that make up Web Services. This ranges from how Web Services compare to other remote object technologies to the underlying XML standards that comprise Web Services such as WSDL, *SOAP*, and UDDI. The general concepts of XML are covered in Chapter 2 in order to give you a quick reference for the terms used in later chapters.

1

Introduction to Web Services

In This Chapter

- Defining Objects and Web Services
- Underlying Web Service Technologies
- Different Web Service Implementations

Microsoft, IBM, and Sun Microsystems are all pushing Web Services as the next great technology to allow developers to create remote objects easily. Earlier remote object technologies, such as COM+ and CORBA, were difficult to implement and had high maintenance costs. Additionally, in the case of CORBA, it was expensive to purchase the operational license. The promise of Web Services is to finally make remote objects a reality, but many of the details, such as security, seem to be hidden or spread across several different Web sites. Plus the term *Web Services* is generic and doesn't hint at all the underlying technologies. This chapter describes at a high level the tool and standards that make Web Services possible and the related technologies.

Web Services encapsulate *Remote Procedure Call* (RPC) with XML as the data packaging. The design of Web Services considers the pitfalls of the aforementioned remote object technologies and tries to avoid them. Although the use of Web Services does solve many of the problems, it also creates many new problems. Such problems are discussed throughout the book.

But before jumping into Web Services, it is important that you understand several basic concepts such as objects, libraries, classes, remote objects, and the like. By understanding these basic ideas, the evolution of Web services will make more sense, and you'll begin to see why the use of Web services has moved to the forefront of Web technology.

DEFINING OBJECTS AND WEB SERVICES

The introduction of object-oriented programming promised the reusability of code across multiple systems and architectures. Many predicted that programmers eventually would not need to create their own objects because code repositories would possess all the necessary code in class files.

Some vendors took advantage of this and provided class libraries that took care of basic functionality. For example, many commercial libraries in C++ contain standard ways of dealing with date and time. This saves the developer time because the date and time functionality is already written in a standard way. When newer languages emerged, such as C# and Java, this basic functionality came as part of the language.

However, the concept of a central repository does not work in every situation. For example, because business logic varies from corporation to corporation, it would be impossible for a repository to contain that code. In this case, commercial libraries can only help by providing basic and generic classes that are commonly needed, but these libraries cannot always contain the necessary logic that a corporation needs because its situation is so unique. Certain business situations, such as in the banking industry, allow vendors to create prepackaged libraries because government regulations enforce certain logic.

At first, objects were only available to programs running on the same machine. For example, on the *Windows* directory of a PC there are several *Dynamic Link Libraries* (dll) files that contain information for *Windows* and other programs to operate. If you do a search for `*.dll` from your *Windows Explorer* in the *Windows* directory, your results may look like Figure 1.1.

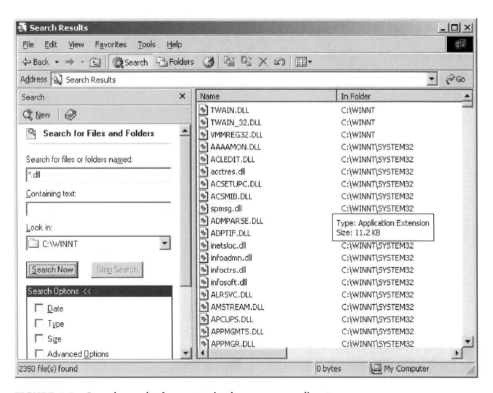

FIGURE 1.1 Search results for `*.dll` in the `c:\WINNT` directory.

Each of these dlls contains information that several applications use to create objects. The dlls reside in the *Windows* directory so applications can find them without much work and different applications can then share the same dll. To get a better idea of a library, class, and object, take a look at the C# example in the following section.

Libraries, Classes, and Objects

The best way to understand *libraries*, *classes*, and *objects* is to see them in action. In the first code example, C# is used to define a library. It is a library because it does not contain a main method for the code to execute, and because the type of project chosen is within *Visual Studio*™.

This library example defines the namespace `HelloWorld`. This is just a unique identifier that prevents the collision of method and class names, which is helpful in large projects. Next, the code defines the class `MyHelloWorld`. A class is a container that possesses all the functionality and values that a particular object needs to contain. Finally, the library defines the method, also known as a function, `DisplayHelloWorld`. This is the only functionality for this class other than the default constructor. Thus, any object instantiated from this class will only be able to call `DisplayHelloWorld`.

```
using System;

namespace HelloWorld
{
  //simple class to display Hello
  public class MyHelloWorld
  {
    public void DisplayHelloWorld()
    {
      Console.WriteLine("Hello World!");
    }
  }
}
```

In the next C# code example, the code, once compiled into an executable, calls the functionality in the library just created. When you compile the previous code example, it creates a dll. When you start a new project to use the code in the following example, be sure to open the *Solution Explorer* window and click on "References." Then browse to the location of the dll created when you compiled the previous library. Figure 1.2 shows the `HelloWorld` dll added to the "References" in the *Solution Explorer*.

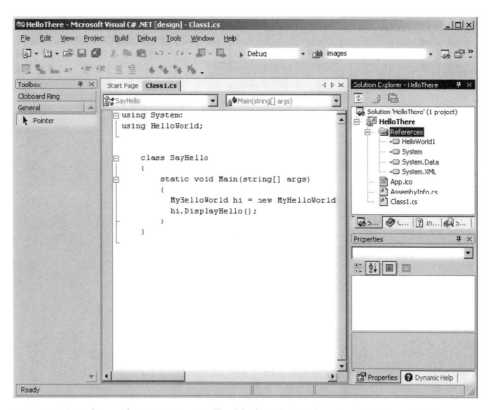

FIGURE 1.2   Shows the `HelloWorld` dll added to the "References" of the project.

Once *Visual Studio.NET* knows where to find the dll, you need to tell your executable that you want to use the functionality in the library. In the next C# example, the code must first define which namespaces it uses. Remember that the previous code example defined the namespace `HelloWorld`. By stating `using HelloWorld;` in the code, the executable knows to use the methods contained in the library example.

The example now defines the class `SayHello` and defines a main method. The main method is an important difference between a library and an executable. The library only contains code for executables and other libraries to use. The executable actually runs.

Within the main method we actually define the object `hi` which is of type `My-HelloWorld`. Remember that `MyHelloWorld` is the name of the class defined in the library. The object hi now calls the only method available to it, which is Display-Hello().

```
using System;
using HelloWorld;


class SayHello
{
  static void Main()
  {
    MyHelloWorld hi = new MyHelloWorld();
    hi.DisplayHello();
  }
}
```

Figure 1.3 shows the executable displaying the output defined in the library example.

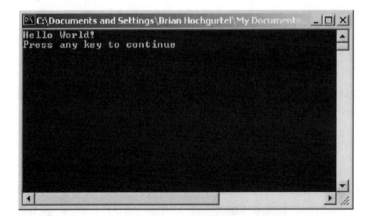

FIGURE 1.3   SayHello executable displaying the output of the DisplayHello method defined in the library example.

Using Remote Objects

Using objects in software is a great way of reusing code at the application level on a local machine. If you're in a corporation and you're maintaining several hundred or thousand workstations and the applications on each system, having these libraries on every system becomes a maintenance headache. If an update is needed, every system within a corporation needs the new code installed. Remote objects, which are objects instantiated on a central server and can be accessed throughout a network including the Internet, become critical at this point.

A remote object is available to a program across a network or even the Internet. *Common Object Manifest* (COM), *Common Object Request Brokerage* (CORBA), *Remote Method Invocation* (RMI), and *Remote Procedure Call* (RPC) all invoke objects remotely and are described in greater detail later in the chapter. The promise of each of these methods allows a developer to change code in one place and then all systems and applications using these objects instantly have access to the new code. The disadvantage here is that if you make a coding error, all the systems now access that error. This causes all the applications using that code to fail. Figure 1.4 illustrates how a remote object can be shared across multiple applications.

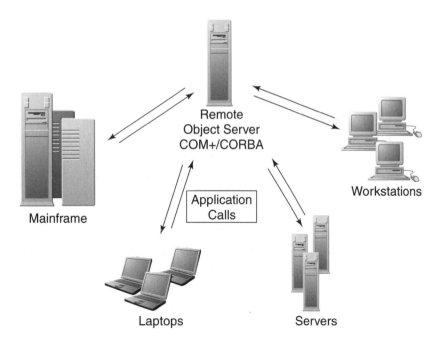

FIGURE 1.4   Remote objects can be shared by several applications in a business setting. In this case, the remote object server contains objects for applications running on workstations, mainframes, laptops, and other servers.

Reaching the promise of remote objects has been difficult. Implementing CORBA takes a great deal of effort and a company incurs great expense to purchase the *Object Request Broker* (ORB). An ORB is the server needed for applications to use remote objects. Only a small percentage of programmers implement CORBA on a daily basis, so it is also difficult for a manager to hire someone who understands the technology.

COM and COM+ offer a somewhat simpler solution to remote object computing. Most of the software needed, such as *Visual Studio* and Microsoft *Transaction Server*, are relatively inexpensive and easily obtainable, but COM and COM+ are typically only available to *Windows* applications. So these objects are not available to applications running under UNIX or another non-Microsoft operating system. In addition, COM takes a highly skilled programmer who understands everything about this technology. Again, it is difficult to find a programmer at this level.

The quest for remote object technologies that are simpler to use and implement brought RMI and RPC into the forefront briefly. They are simply request and response systems whose function is similar to how a Web page responds to a request from a browser. The problem here lies with the implementation that each vendor decided to pursue, and thus working with RPC software from one vendor wasn't always compatible with another's. Again not having cross-platform compatibility prevented remote object technology's general acceptance.

WHERE WEB SERVICES AND *SOAP* FIT IN

In the late 1990s, Microsoft realized the weaknesses in its own remote object technology, COM and COM+. As they begun their *.NET* initiative they saw that they needed something new in order for remote object technology to succeed.

One of the aspects of remote objects Microsoft needed for success was the ability for the technology to be cross platform, and because all of Microsoft's products run under their own operating system, they needed a creative solution. They created an XML standard called the *Simple Object Access Protocol* (*SOAP*) which is the combination of an *eXtensible Markup Language* (XML) document and a standard protocol that can work across the Internet. The protocols *SOAP* uses to transmit data include the *Simple Transport Mail Protocol* (SMTP), the *File Transfer Protocol* (FTP), and the *Hypertext Transfer Protocol* (HTTP). Developers often refer to these as the Web service's protocol stack.

Once Microsoft's team perfected the *SOAP* standard, they open-sourced this technology by giving it to the *World Wide Web Consortium* (W3C). The W3C made *SOAP* an official standard so other vendors could start making similar remote object technologies that were compatible with the Microsoft technology. Web Service products that exist on other platforms, such as UNIX or AS400, are instantly compatible with Microsoft technologies running under various versions of *Windows*. This happened because Microsoft chose standard technologies such as XML and the various protocols that exist on all platforms.

Currently, with the release of *.NET*, many vendors seek to create Web Service products that work with Microsoft's. Soon it won't matter what platform or language you are using. With Web Services, remote objecting may finally reach its promise because Web Services are simple and standard and vendors are making them easier to implement.

Before moving on, it is important to realize the difference between *SOAP* and a Web Service. *SOAP* is the actual protocol for moving data across the Internet. A Web Service is an object that uses *SOAP* to transmit data to an application or Web page. For example, a Web Service provides stock quotes whereas *SOAP* is that standard that allows the stock quote Web Service to be compatible across platforms.

UNDERLYING WEB SERVICE TECHNOLOGIES

As with any XML technology, there are underlying technologies that make the overall concept of Web Services work.

Remember that a Web Service is the actual node out on the Internet that performs some sort of function and then returns data.

The underlying technologies that allow you to use a Web Service include the protocol that moves the data across the Internet or network (such as HTTP) and *SOAP*. Other technologies that allow you to use and discover Web Services include *Web Service Description Language* (WSDL) and *Universal Discovery, Description, and Integration* (UDDI). This section briefly introduces these standards and chapters later in this book go into greater detail.

SOAP and XMLP

SOAP, as mentioned earlier in this chapter, is the underlying XML standard that allows Web Services to work cross platform because the data is transmitted in a standard way. *SOAP* is a constantly evolving standard; soon it will no longer be known as *SOAP* but as the *XML Protocol* (XMLP). Chapter 3 goes into great detail about the *SOAP* standard, its terminology, and syntax.

UDDI and Discovery

Communication about the location, needed parameters, necessary tools, and other information about a set of remote objects turned out to be another difficulty when developers deployed these technologies. There was no standard way to describe the objects or location to put the description. Usually a document created by developers circulates internally with all the classes and methods available. This works great within the corporate infrastructure.

If these objects need to be shared across the Internet with thousands of users, the documentation will need to reach them. This can be done with a Web page, but somehow the users would need to discover this information. UDDI provides a standard way to discover and describe Web Services without human intervention.

UDDI appears to have two roles. In one of its roles, it's an XML standard that describes where to find particular Web Services. In addition, there are several Web sites that use UDDI to advertise available Web Services. Some sites to look at for Web Services include *www.uddi.org*, *www.xmethods.com*, and *uddi.microsoft.com*. Chapter 4 discusses UDDI and discovery in more detail.

DESCRIBING A PARTICULAR WEB SERVICE

When a client such as an application or a Web page uses a Web Service, this is know as *consuming* it. Web Services Description Language (WSDL) describes the Web Service to the client so it knows what classes and methods are available along with the location on the Internet. WSDL is another XML-based language.

WSDL is especially important with technologies from Microsoft. A WSDL file must exist for a Microsoft Web Service to consume any methods. With some of the Java technologies, it is possible to sometimes use a Web Service by just knowing the URL of the Web Services. Chapter 4 explores the syntax and many uses of WSDL.

DIFFERENT WEB SERVICE IMPLEMENTATIONS

There appears to be two major technological camps in the Web Services industry. The first camp is Microsoft. Microsoft got a head start because its people developed the *SOAP* standard and then gave it to the open-source community. So before other developers knew about the *SOAP* standard, Microsoft had already begun developing programming languages such as C# and *Visual Basic.NET*™ to create a proprietary implementation.

The second camp in the Web Services industry revolves around Java™, but it isn't only Sun Microsystems, the creator of Java, deploying technologies. In this case, several vendors are each plying for a piece of the marketplace. Some of these vendors include BEA, Cape Clear Software, IBM, and many more. In this book, the focus is based on servers and Web Service libraries that are free and easily downloaded by the reader. This includes technology from Sun and the Apache group.

The important thing to remember is that it's ok for there to be several vendors supporting Web Services. Unlike previous technologies, such as RPC, the underlying technologies are based on XML standards. So even though the functionality may be different, the use of these standards almost ensures compatibility across platforms.

Microsoft Implementation

As mentioned earlier in this section, Microsoft created a head start for itself by creating the *SOAP* standard; thus, its Web Services software is probably the easiest to use and deploy. Currently this technology is only available on the *Windows* platform with some rumblings of the *.NET* platform; therefore, Web Services will be available on other platforms such as Linux.

The advantages of Web Services under the umbrella of Microsoft's *.NET* platform are ease of use and discovery. Microsoft provides a large number of tools to make the creation and use of Web Services very easy. This includes the automatic generation of WSDL, discovery tools such as disco (which searches servers that have *.NET* Web services), browser-based testing and discovery of methods, and easy creation of Web Services within Microsoft's proprietary languages. In fact, it is just a matter of adding a few lines of information, not necessarily code, which makes an existing method a Web Service.

Examples in Chapter 6 show in greater detail all these available features.

Although easy to use and deploy, all of the Microsoft Web Service technologies rely on their Web server *Internet Information Server* (IIS). While this is probably the easiest Web server to maintain and configure, it has also had its share of security problems, including the Code Red Virus. Therefore, more maintenance may be needed with IIS because you'll need to track any security patches Microsoft releases.

Java Implementation

The Java world is definitely catching up to Microsoft when it comes to Web Services. Nothing has quite reached the ease of use of Web Services under *.NET*. Because Microsoft developers were part of the creating the *SOAP* standard, they had a sneak peak to upcoming technologies. In addition, the Java world tends to ignore any standard that Microsoft supports. This is the political quagmire that has existed in the IT world for years.

As the Java world sees the reaction to Web Services under *.NET*, more and more Web Services products start showing up. BEA, Sun, and IBM are starting to deliver more and more Web Service products. In addition, the Apache group provides a great free *SOAP* library to access Web Services. Thus, Java-based Web Services are starting to catch up.

There are a lot of advantages to using Web Services with Java. First of all, several different vendors implement Web Services with Java. This gives you greater selection of products to choose from during implementation, and perhaps allows you to easily integrate Web Services into your already existing architecture.

Java Web Services work on top of both Java *Server Pages* (JSP) and servlets, and a developer's choice for serving these technologies is far greater than with Microsoft's implementation. *Tomcat*™, which is a free Java server that integrates easily with *Apache*™, is free from the *Apache* group. IBM's *Websphere*™, BEA's *WebLogic*™, and Sun's *iPlanet*™ server are all commercially available Web servers that allow a developer more options when deploying Web Services. This is unlike Microsoft's implementation where you are locked into one platform and one Web server.

OTHER TECHNOLOGIES

Java and *.NET* are not the only technologies that need Web Services. There are third-party products available that allow Web Services to integrate with CORBA, COBOL, C++, and other legacy systems. By using Web Services with these systems, a corporation saves money because legacy systems don't need to be replaced. Instead, by adding a Web Services layer on top of these systems, the information these systems contain is available to applications that consume Web Services.

For example, if a corporation such as a bank utilizes a lot of COBOL and needs to have their Web site communicate with these systems, a Web service layer on top of the systems allows this type of communication. The cost of the software is minimal compared to replacing large legacy implementations such as a billing system at a large telecom or transaction software at a large bank. Figure 1.5 illustrates how a Web Services layer on top of a legacy COBOL banking system can bring data from a legacy COBOL banking system to a .NET- (or Java-) based Web site.

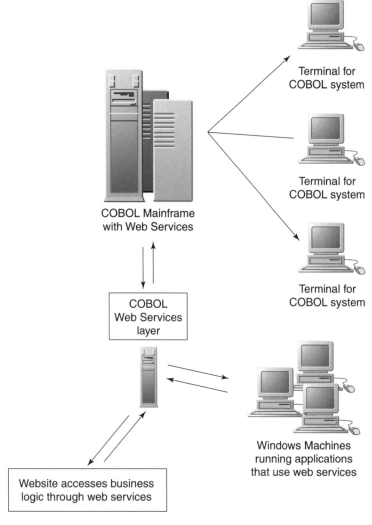

FIGURE 1.5 How Web Services may bring data from a legacy COBOL banking system to a *.NET-* (or Java-) based Web site.

CONCLUSION

This chapter briefly covers many of the aspects involved with Web Services and some of the history of Web Services. It's important to see how all the different technologies fit together, and why it's important to study cross-platform Web Services. As more and more corporations either merge or set-up *Business to Business* (B2B) exchanges, it will be important to understand how to integrate divergent systems such as Java and *.NET*.

Understanding the technologies that came before Web services, such as CORBA and COM+, will help you understand why there is such a push for Web Services. It is a standardized way to call objects across platforms because it uses XML to transmit and receive data. You'll see how XML is used when WSDL, *SOAP*, and UDDI are discussed in the following chapters.

2 Brief Introduction to XML

In This Chapter

- Building an XML Document
- Well-Formed and Valid XML
- Self-Describing XML

XML lies under the hood of Web Service products. It's present but the programmer rarely works with it directly unless creating a new product in a language that doesn't already support Web Services. It is often possible to look at the XML a Web Service uses to communicate. Therefore, it's important for a programmer to understand some basic XML to work with Web Services.

In addition, many of the underlying technologies for Web Services, such as *SOAP*, use XML. So regardless of how high level a Web Service technology may be, it is useful to know what is going on underneath the hood.

A book on Web Services cannot cover the entire concept of XML in a single chapter, but in this short chapter several of the concepts needed to understand the XML in *SOAP*, UDDI, and WSL are covered. For a complete introduction, take a look at the actual standard at the *World Wide Web Consortium* (W3C) at *http://www.w3.org/XML/*.

BUILDING AN XML DOCUMENT

To read an XML document, an application uses a *parser* to get the data contained in the document. A parser usually consists of a large API that allows the programmer to choose which elements to look at in the document. Microsoft's *.NET* architecture provides a developer with several classes for accessing the data in a document, and the *Apache* group develops a parser called *Xerces*™ that works cross platform.

With Web Services, an application passes an XML document across the Internet with different transport protocols. Therefore, either a client or sever side program must parse the XML to get to the data within the document. The following sections describe many parts of an XML document that a parser encounters.

Processing Instruction

The first part of any XML document is the *Processing Instruction* (PI). This tells the parser that the data is in an XML document and the version of XML used (at this point it's always 1). The start of the document now looks like the following.

```
<?xml version="1.0" ?>
```

The version is always set to 1 because there hasn't been another version of XML. This statement tells the parser where to begin looking for XML.

Root Element

To have a useful document, data needs to be present. To begin describing data, a root element must be present. This is the outermost element in the document. An element is simply a tag that looks much like an HTML tag, but in the case of XML the programmer chooses the name of the tag. For this example, BOOK is the root element.

```
<?xml version="1.0" ?>
<BOOK>
</BOOK>
```

The element is the word BOOK surrounded by <>. The element with the slash, in this case </BOOK> is the closing element. An XML document must have only one root element, and this element must be the outermost element.

Later in the book you'll see that the root element begins the definition of a *SOAP* document or a WSDL file.

Empty Elements

With the small amount of data present in this document, the closing element isn't really necessary. Using an *empty element*, which is an element with no closing tag, makes the data more succinct by just using a / at the end. If we take the previous example and make BOOK an empty element, we have the following result.

```
<?xml version="1.0" ?>
<BOOK TITLE="Cross Platform Web Services"/>
```

Attributes

Additional information added to an element is an *attribute* that, in this case, is part of the opening BOOK element and it contains the title of a book. Attributes always appear as part of the opening element and can be in any element in the document (not just in the root element as in the examples thus far).

```
<?xml version="1.0" ?>
<BOOK TITLE="Cross Platform Web Services">
</BOOK>
```

The XML standard contains a great deal of flexibility because both elements and attributes are allowed. This gives you and the developers of an XML language, such as *SOAP*, great flexibility in design.

As shown in examples later in the chapter, attributes often define namespaces or locations, such as the next *SOAP* node, for the XML document. Be sure to see the definition of namespaces later in this chapter.

Attribute Centric Data

So far this document doesn't really give a user much information about the book. By adding more attributes to the document, a better definition of data occurs and the following is the possible result.

```
<?xml version="1.0" ?>
<BOOK TITLE="Cross Platform Web Services"
      PAGECOUNT="400"
      AUTHOR="Brian Hochgurtel"
      PUBLISHER="Charles River Media"/>
```

This document is attribute centric because the information all resides within attributes.

Element Centric Data

Now the information is more descriptive, but another possibility it to format the data which elements that are children of BOOK, such as the following.

```
<?xml version="1.0" ?>
<BOOK>
  <TITLE>Cross Platform Web Services</TITLE>
  <PAGECOUNT>400</PAGECOUNT>
  <AUTHOR>Brian Hochgurtel</AUTHOR>
  <PUBLISHER>Charles River Media</PUBLISHER>
</BOOK>
```

Because the data in this example belongs completely in elements, the document is considered element centric.

Elements and Attributes in the Same Document

This example is *element centric* because all the data resides in elements and the previous example was *attribute centric* because the data resides in attributes. XML, however, does not require that a document be attribute or element centric because the data can be mixed, as shown in the following example.

```
<?xml version="1.0" ?>
<BOOK TITLE="Cross Platform Web Services">
  <PAGECOUNT>400</PAGECOUNT>
  <AUTHOR>Brian Hochgurtel</AUTHOR>
  <PUBLISHER>Charles River Media</PUBLISHER>
</BOOK>
```

Nested Elements

The elements chosen for this document only allow for one book to be in the document. If several books need to be in the document, more nesting needs to occur. By nesting BOOK under a different root element named LIBRARY, several occurrences of BOOK can occur in the document, as shown in the following example.

```
<?xml version="1.0" ?>
<LIBRARY>
  <BOOK TITLE="Cross Platform Web Services">
    <PAGECOUNT>500</PAGECOUNT>
    <AUTHOR>Brian Hochgurtel</AUTHOR>
    <PUBLISHER>Charles River Media</PUBLISHER>
  </BOOK>
  <BOOK
    TITLE="Learning Visual Basic Through Applications">
    <PAGECOUNT>418</PAGECOUNT>
    <AUTHOR>Clayton E. Crooks II</AUTHOR>
    <PUBLISHER>Charles River Media</PUBLISHER>
  </BOOK>
</LIBRARY>
```

LIBRARY is now the root element. BOOK is a child of LIBRARY but is still the parent of PAGECOUNT, AUTHOR, and PUBLISHER. This document now has the ability to describe multiple books, and perhaps other items, that may fit within a LIBRARY such as a magazine.

Using Namespaces

Namespaces ensure that the element names used in your XML document are unique, and have many of the same properties as the namespaces used in the C# example shown earlier in the chapter.

The namespace definition occurs in the root element (the outermost element) and utilizes a URL as a unique identifier. Realize that there is no required content at the URL. It's just an identifier that assists in making the elements unique. Defining the namespace occurs in the root element, as the following example illustrates.

```
<BOOK XMLNS:WEBSERVICES="www.advocatemedia.com/XML"></BOOK>
```

Then all the child elements of BOOK begin with the namespace.

```
<?xml version="1.0" ?>
<BOOK XMLNS:WEBSERVICES="www.advocatemedia.com/XML">
  <WEBSERVICES:TITLE>Cross Platform Web Services</
    WEBSERVICES:TITLE>
  <WEBSERVICES:PAGECOUNT>400</WEBSERVICES:PAGECOUNT>
  <WEBSERVICES:AUTHOR>Brian Hochgurtel</WEBSERVICES:AUTHOR>
  <WEBSERVICES:PUBLISHER>Charles River Media</
WEBSERVICES:PUBLISHER>
  </BOOK>
```

Namespaces are an important concept to understand because many of the XML standards underlying Web Services utilize them usually as a way to represent various elements that are vendor dependent or to support primitive types from schemas.

WELL-FORMED AND VALID XML

An XML document is often referred to as being *well-formed* and *valid*. This means that the document meets all the rules discussed in the previous section and contains all the information specified in either a *Document Type Definition* (DTD) or in an XML schema. Both act as a packing slip for XML documents, specifying which data needs to be present in the document. Validating a document against a schema or a DTD is a costly process and, thus, probably only occurs during the development of *SOAP* software. Once a developer ensures that his software produces the correct XML in the *SOAP* transactions, the validation is probably turned off. This is all dependent on how each vendor implements Web Services.

Well-Formed XML

A well-formed XML document follows the rules set forth by the W3C. Put simply, there must be one or more elements, there can only be one root element that is not overlapped by any other element, and every start tag must have an end tag unless it's an empty element. Thus, one of the original examples was valid when it just had one empty element.

```
<?xml version="1.0" ?>
<BOOK TITLE="Cross Platform Web Services"/>
```

By removing the / at the end of the element, the BOOK element no longer has a closing / or element. Therefore, the following document is not well formed.

```
<?xml version="1.0" ?>
<BOOK TITLE="Cross Platform Web Services">
```

But by adding a closing BOOK element, the document becomes well formed again.

```
<?xml version="1.0" ?>
<BOOK TITLE="Cross Platform Web Services"></BOOK>
```

Another error that prevents a document from being well formed happens when the root element gets overlapped by another tag. In the following example, BOOKDATA overlaps the root element BOOK and this causes a parsing error.

```
<?xml version="1.0" ?>
<BOOK TITLE="Cross Platform Web Services">
 <AUTHOR>Brian Hochgurtel</AUTHOR>
 <BOOKDATA>
    <PAGECOUNT>400</PAGECOUNT>
    <PUBLISHER>Charles River Media</PUBLISHER>
 </BOOK>
 </BOOKDATA>
```

A quick way of checking the well-formedness of a document is to have Internet *Explorer* view the document. Figure 2.1 shows how Internet *Explorer* reports the error in the previous example.

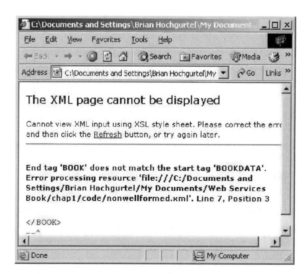

FIGURE 2.1   Using Internet *Explorer* to report well-formedness errors in XML documents.

Validity and Document Type Definition (DTD)

DTDs are a hold over from the older *Serialized General Markup Language* (SGML) standard that the publishing industry created to publish books. They are slowly falling out of favor with developers because they do not use XML. Up until the recent time, DTDs were the only way to have a valid XML document, but they didn't provide many of the things needed for common programming such as types or order. They did, however, allow a user to specify entity references that gave the ability to substitute values in and out of XML documents. Consider the following simple XML document.

```
<?xml version="1.0" ?>
<BOOK TITLE="Cross Platform Web Services">
    <PAGECOUNT>400</PAGECOUNT>
    <AUTHOR>Brian Hochgurtel</AUTHOR>
    <PUBLISHER>Charles River Media</PUBLISHER>
</BOOK>
```

NOTE

DTDs rarely occur outside the publishing industry. However, you will find that many tools, such as Sun Microsystems Forte™, *allow you to easily create them.*

A simple DTD for this file would need to recognize that BOOK is the root element and TITLE, PAGECOUNT, AUTHOR, and PUBLISHER are all children of BOOK. Because BOOK is a root element, it cannot be optional but all the other elements can be. We also need the DTD to recognize that TITLE is an attribute of BOOK. The following is the appropriate DTD for this XML document.

```
<?xml version="1.0" ?>
<!-- This is a comment -->
<!-- The following code is the DTD -->
<!-- The PI and the DTD are the prolog of the document
-->
<!DOCTYPE BOOK [
<!ELEMENT BOOK (PAGECOUNT?,AUTHOR+,PUBLISHER+)>
<!ATTLIST BOOK TITLE CDATA #REQUIRED>
<!ELEMENT PAGECOUNT (#PCDATA)>
<!ELEMENT AUTHOR (#PCDATA)>
<!ELEMENT PUBLISHER (#PCDATA)>
]>
<BOOK TITLE="Cross Platform Web Services">
  <PAGECOUNT>400</PAGECOUNT>
  <AUTHOR>Brian Hochgurtel</AUTHOR>
  <PUBLISHER>Charles River Media</PUBLISHER>
</BOOK>
```

The DTD at the beginning of the document is clearly not XML. It's a completely different language and doesn't provide many of the constructs needed to be useful to a developer. For example, it is not possible to specify the minimum or maximum number of AUTHORS that need to be in the document. Specifying quantities is done with the symbols +, *, and ?. The + means 1 or more of the element whereas the * means 0 or more. The ? means the element is optional.

The vagueness and the difficult syntax of DTDs cause most developers to look at schemas as a way to validate XML documents.

Validity and Schemas

The following XML code example is a schema generated by *Visual Studio.NET*. In the code, there are a lot of namespaces defined and many of them deal specifically with things *Visual Studio* needs, but there is still important information present in this code that helps a developer more than any DTD could.

Skip past the namespace definitions and look at the first xs:element. The first xs:element defines the requirement for PAGECOUNT in the XML document, and these requirements are that it is a string according to the type attribute, minOccurs set to 0 indicates that PAGECOUNT is not required, and maxOccurs means that PAGECOUNT can appear a maximum of three times. Additionally, the xs:sequence tag allows you to determine the order of the elements in the schema.

```xml
<?xml version="1.0" ?>
<xs:schema id="NewDataSet"
  targetNamespace="http://www.advocatemedia.com/~vs1CO.xsd"
  xmlns:mstns="http://www.advocatemedia.com/~vs1CO.xsd"
  xmlns="http://www.advocatemedia.com/~vs1CO.xsd"
  xmlns:xs="http://www.w3.org/2001/XMLSchema"
  xmlns:msdata="urn:schemas-microsoft-com:xml-msdata"
  attributeFormDefault="qualified"
  elementFormDefault="qualified">
<xs:element name="BOOK">
  <xs:complexType>
    <xs:sequence>
    <xs:element name="PAGECOUNT" type="xs:string" minOccurs="0"
      maxOccurs="3" msdata:Ordinal="0" />
    <xs:element name="AUTHOR" type="xs:string" minOccurs="0"
      msdata:Ordinal="1" />
    <xs:element name="PUBLISHER" type="xs:string" minOccurs="0"
      msdata:Ordinal="2" />
    </xs:sequence>
    <xs:attribute name="TITLE" form="unqualified" type="xs:string"
      />
  </xs:complexType>
</xs:element>
<xs:element name="NewDataSet" msdata:IsDataSet="true"
      msdata:EnforceConstraints="False">
  <xs:complexType>
    <xs:choice maxOccurs="unbounded">
      <xs:element ref="BOOK" />
    </xs:choice>
  </xs:complexType>
</xs:element>
</xs:schema>
```

Schema types are often used to identify primitive types in a SOAP message.

NOTE

The XML in a schema is quite complex and confusing, but rarely does a programmer need to worry about coding a schema by hand. There are several tools available, such as *Visual Studio.NET*, that generate schemas automatically based on a given XML document. The most a programmer might do is go in and modify any defaults such as `minOccurs` and `maxOccurs`.

It is important to notice that the schema does represent the primitive type that the value of the XML element must appear as. In this case, all the values comprise of `xs:string` but other types may be represent as well such as `xs:int` or `xs:bool`. Schema is one of the few XML standards that represent types in this manner.

Most developers favor schemas in software development. They provide many definitions, such as type and order, needed to generate code or to validate technical XML documents.

SELF-DESCRIBING XML

Consider the following data.

```
Cross Platform Web Services Using C# and Java, Brian
Hochgurtel, 500, 2003
```

Before the advent of XML, transmitted data often looked like the previous data. A programmer would rely on another individual to state which field represented which data. The elements used in XML documents should describe the data they represent. When we covert the simple text data shown previously into XML, the results may look like the following.

```
<?XML VERSION="1.0"?>
<BOOK>
  <TITLE>Cross Platform Web Services</TITLE>
  <AUTHOR>Brian Hochgurtel</AUTHOR>
  <PAGECOUNT>500</PAGECOUNT>
  <YEARPUBLISHED>2003</YEARPUBLISHED>
</BOOK>
```

In this example, the elements, such as `<TITLE></TITLE>`, describe the data represented. The XML standard allows a programmer to choose any element name needed to describe the data, but if the names chosen aren't descriptive, the following could be the result.

```
<?XML VERSION="1.0"?>
<THING>
  <NAME>Cross Platform Web Services</NAME>
  <PERSON>Brian Hochgurtel</PERSON>
  <NUMBER>500</NUMBER>
  <YEAR>2003</YEAR>
</THING>
```

You no longer know that the data represents a book. Instead, it represents some sort of thing but you are no longer sure what type of thing it represents. In the previous example, you knew that the data represented information for a particular book.

You will see that *SOAP* messages are not really self-describing. Remember that these messages are meant for applications to read, so the message formats are not as clean as some of the other XML languages such as the *eXtensible Stylesheet Language* (XSL).

CONCLUSION

In this chapter, several concepts of XML that are relevant to Web Services were introduced. The most import concepts are the ideas of primitive schema types, namespaces, elements and attributes, and the concept of a well-formed XML document. These are terms that will reappear in the next few chapters as you explore *SOAP*, WSDL, and UDDI. By reading this chapter, several of the key XML concepts should be fresh in your mind.

3

SOAP

In This Chapter

■   Vocabulary
■   Putting the Vocabulary into Action
■   The *SOAP* XML Document
■   *SOAP* Documents and Transport Bindings

S*OAP* originally stood for the "Simple Object Access Protocol," but with the latest revision, *SOAP* is no longer an acronym. This reflects the dynamic nature of the *SOAP* standards and Web Services, especially as various implementations start to mature.

The *SOAP* standard dictates how the XML looks within the *SOAP* document, how the contents of that message are transmitted, and how that message is handled at both the sender and receiver. *SOAP* also provides a standard set of vocabulary. This chapter introduces that vocabulary, shows different applications of that vocabulary, and shows what the XML in a *SOAP* document looks like.

VOCABULARY

As with any technology, *SOAP* comes with its own set of vocabulary. There are several terms used frequently to describe different aspects of the *SOAP* standard. You'll find that many developers use these terms without truly understanding their meaning. So it's important to take the time to understand what each term means and how it applies to both the *SOAP* standard and to an actual Web Service.

The SOAP *standard is not just an XML standard. The standard includes how* SOAP *messages should behave, the different transports used, how errors get handled, and more.*

The terminology related to *SOAP* comes in two different categories: transmission and message. The terms related to transmission deal with describing how *SOAP* messages relate to the protocol, how the different *SOAP* nodes converse, and so on. On the other hand, terms related to the XML within *SOAP* fall into the message category.

Transmission

In the next set of terms, the focus is strictly on describing the transmission of *SOAP* messages between the various senders and receivers. The "*SOAP* Connection Model" section of this chapter actually demonstrates the terms defined here with graphics.

SOAP

This used to stand for "Simple Object Access Protocol" but with the latest revision of the standard, Version 1.2, *SOAP* is no longer an acronym. The *SOAP* standard contains the information for how the messages should be sent, the format the XML appears in, the different primitive types supposed, the roles different pieces of software take during the transmission of the *SOAP* documents, and the type of transports available, such as HTTP.

As this standard moves forward, the name will change to XMLP for the XML Protocol. *SOAP*'s evolution has been muddied by the fact that it came from Microsoft. XMLP is a complete rewrite of the standard so that Web Services become even more cross-platform compatible. Because of this change, much of this vocabulary must evolve. Although the names will probably change, the general idea of each term will remain the same.

SOAP Binding

This describes how a *SOAP* message works with a transport protocol such as HTTP, SMTP, or FTP to move across the Internet. It is important that *SOAP* moves across a standard protocol in order to communicate with other Web Service products.

Before *SOAP*, many developers created their own method of transmitting XML documents through a network. This works fine as long as the transmission is limited within a particular team. If, however, you need to work with another group either within or outside your company this becomes difficult because of training and possible modification to work with an XML transmission they may be using. By using a standard XML document on standard protocols, the work needed for collaboration will be minimal.

SOAP Message Exchange Pattern (MEP)

This describes how a *SOAP* document gets exchanged between a client and server. The *SOAP* message possesses a binding, such as HTTP, so that it can move across the Internet. The conversation between the client and server, both known as nodes, determines what actions both take. See the next section for some simple diagrams that help show how this works.

Remember that *SOAP* is an XML encapsulation of RPC. Therefore, the MEP is completely request and response between the client and server (or other nodes). Thus, if there needs to be several interactions between nodes, this takes several requests and responses to complete transmission. This differs from other remote object technologies such as CORBA where the entire conversation occurs over one single connection.

SOAP Application

A *SOAP* application is simply an application that uses *SOAP* in some way. Some applications maybe entirely based on the *SOAP* standard, such as the stock Web Services example shown later in the chapter, or may just use the *SOAP* standard to receive code or software updates. Remember an application can produce, consume, or be an intermediary (or router).

SOAP Node

A node's responsibility can include sending, receiving, processing, or retransmitting a *SOAP* message. A node is just a piece of software that properly handles a *SOAP* document dependent on its role. Besides transmission, a node is also responsible for enforcing that the XML contained in the *SOAP* document is grammatically correct according to the *SOAP* standard.

SOAP Role

A *SOAP* role defines what a particular node does. It may be a sender, receiver, or intermediary.

SOAP Sender

The node sending the *SOAP* request is the *SOAP* sender. If you think of a client/server example, when a client first makes a request, it sends a message to the server asking for some information. In the upcoming Stock Quote Example, the client sends a request to the Stock Quote Server. In this case, the client acts as the *SOAP* sender by transmitting a message asking for a quote.

SOAP Receiver

A server that receives the *SOAP* request is obviously the receiver. This is the server in the client/server model. The Stock Quote Example later on in the chapter has a server that receives the request for stock quotes and then returns the appropriate values.

SOAP Intermediary

An intermediary looks at a *SOAP* message, perhaps acts on some of the information in the message, and then looks at the *SOAP* document for more information on where to pass the information in the document next.

A *SOAP* intermediary essentially acts like a router in a network. A router takes a look at a packet of information moving through a network, finds the packet's next destination, and then sends it to that destination.

A *SOAP* intermediary does the same thing but it's looking at *SOAP* messages and information in the XML to send the message to the proper location. This occurs when a large corporation possesses many *SOAP* servers that perform different functions, and the *SOAP* intermediary may have access to the firewall. Once it receives the information, it looks at the XML to see where to send the message next. It may act on or modify the data before this retransmission, but it is not necessary.

Message Path

A *SOAP* message moves from sender to receiver perhaps through several intermediaries. The resulting route the message takes is the Message Path.

Initial *SOAP* Sender

The node sending the first *SOAP* request is the initial *SOAP* sender.

SOAP Feature

A *SOAP* feature is a piece of functionality in software supporting *SOAP* that deals with a feature of *SOAP*. Examples include securing the transaction with a secure protocol or the software acting as an intermediary.

Terms Related to the XML

The *SOAP* standard also defines a small set of XML elements to encapsulate the data transmitted between nodes. There are really only a few elements because the body of the message can vary depending on the implementation. This flexibility is allowed by the standard.

SOAP Message

This is the XML document transmitted by either a *SOAP* sender or receiver. A sender or client creates an XML document containing the information the client needs from the server. Once that document is transmitted, the server parses the information in the document to access the various values and then creates a new *SOAP* message as the response.

SOAP Envelope

This is the root element of the *SOAP* XML document. The *SOAP* document contains several namespace definitions but the elements related to the *SOAP* message will have ENV: as the prefix. Examples later in the chapter describe the XML in greater detail.

SOAP Header

The first part of a *SOAP* message contains a header block in XML that is for the routing and processing of the *SOAP* message. This data is separate from the body of the document which has information related to the object call being made.

SOAP Header Block

A *SOAP* header containing several delimited sections or blocks of information has a header block. These header blocks come with processes that include *SOAP* intermediaries because a node needs to know where to send the message to next.

SOAP Body

The body of the message actually contains the information for the object to process the information. The body is still in XML and, once parsed, the information goes to the object. The object processes the information and the result is put into the *SOAP* body of the returned document.

SOAP Fault

This is simply a piece of information in the XML of a *SOAP* document containing information related to any error that may have occurred at one of the *SOAP* nodes.

Quick Reference

As you move forward in this book, you may find it useful to refer back to this chapter to look up the definition of a term you may have forgotten. The following tables help by giving you a quick reference to each one of the terms. Table 3.1 provides a summary of all connection-related terms. Table 3.2 summarizes *SOAP*-related terms.

TABLE 3.1 Provides a Quick Summary of All the Connection-Related Terms

Connection Term	Description
SOAP	The standard defining how the XML document looks and how the information moves across the Internet.
SOAP Binding	Describes how *SOAP* interacts with a standard transport protocol such as HTTP or SMTP.
SOAP message Exchange Pattern (MEP)	This is the conversation a client and server has while exchanging information.
SOAP Application	An application that consumes or creates *SOAP* messages.
SOAP Node	A server that somehow interacts with the *SOAP* message Exchange Pattern.
SOAP Role	A *SOAP* node may have one of three roles or jobs: sender, receiver, and intermediary.
SOAP Sender	A *SOAP* sender is the node sending a request to another node.
SOAP Receiver	A *SOAP* receiver is the final processor or destination of the request.
SOAP Intermediary	Acts as a router or relay by passing the *SOAP* message onto the next node. Relies on information in the message to find the next node.
Message Path	The path or route the *SOAP* message follows during processing.
Initial *SOAP* Sender	The node that originated the *SOAP* request.
SOAP Feature	Relates to pieces of the *SOAP* standard such as security or errors.

TABLE 3.2 A Summary of All the Terms Related to the *SOAP* message

SOAP *message Term*	*Description*
SOAP message	The XML document transmitted between *SOAP* nodes.
SOAP Envelope	The root element of the *SOAP* XML document.
SOAP Header	The top portion of the *SOAP* XML document that contains information relevant to the processing of the message.
SOAP Header Block	If a header contains a lot of information, several section or blocks occur in the header to separate the information.
SOAP Body	The part of the *SOAP* XML document right below the header that contains information for the actual object call.
SOAP Fault	Information in the *SOAP* document relevant to any error that occurred.

PUTTING THE VOCABULARY INTO ACTION

The last section covered a great deal of new terminology. Covering a few examples should lend to greater understanding.

A Simple *SOAP* Transaction

Figure 3.1 shows a simple model of a *SOAP* transaction. The first item to note is the different protocols available for transmission. FTP, HTTP, and SMTP are available. This provides a great deal of flexibility especially if consumers of your Web Service are in an area that doesn't have a good connection to the Web but still has e-mail access.

The *SOAP* message is the XML document being sent to the *SOAP* receiver. The message contains information in XML that is eventually processed by an object on the receiver. The *SOAP* nodes are both the *SOAP* sender and receiver. Think of the *SOAP* sender as a Web page or an application that consumes the Web Service, and the *SOAP* receiver as the server that hosts the Web Service.

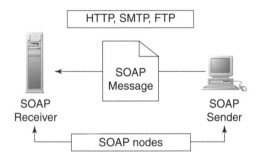

FIGURE 3.1 This image demonstrates a simple *SOAP* transaction.

Using an Intermediary

In the next example, Figure 3.2, a *SOAP* intermediary node is introduced. As mentioned earlier, an intermediary acts as a Web Services router sending requests to the appropriate location based on data found in the XML of the *SOAP* document.

In this case, the *SOAP* sender is a client for a large brokerage house. This brokerage house possesses several *SOAP* receivers each with a different function. By having the *SOAP* intermediary, the brokerage house only needs to publish one address to fulfill all the different requests. Once a *SOAP* message reaches the intermediary, the XML inside the message reveals the next node the request needs to go to. In this case, the brokerage house is offering services for stock quotes, account information, and stock trading. Thus, the XML contains information telling the intermediary where to send the message next. The important thing to note is that the intermediary can act on other information in the *SOAP* message, but it is not required to do so.

How *SOAP* Differs

This book mentions several times that *SOAP* is just a request and response system. This is much like a browser making a request to a Web page. You enter the address into the browser and the request goes out to the Internet and finds the server. Then the server sends the response in the form of HTML; then there is no longer a connection between client and server.

Figure 3.3 shows this request and response pattern between two *SOAP* nodes. This is a conversation between the two nodes about getting some stock quotes. In

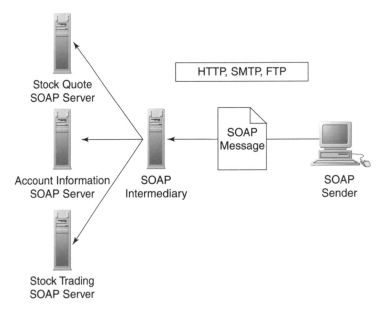

FIGURE 3.2   How a *SOAP* message can be routed by a *SOAP* intermediary.

the first request, the *SOAP* client asks for stock quotes for the symbols: "GE," "C," and "DJI." The response is "35," "48," and "10,000." The *SOAP* client then asks for a quote for the symbol "HOCH," and the receiver sends back information stating there is not such symbol. This pattern of request and response over related information is the message Exchange Pattern mentioned earlier in the chapter.

The difference between a client and server in COM+, CORBA, or another remote object technology is the fact that the entire conversation occurs over a single connection. Figure 3.4 shows the same exchange of information over stock quotes as the previous example In this case, however, there is only one connection, and it is held until the entire conversation is complete. This connection is much more like telnet—a session is held constant between a client and server. Because the connection is constant, the response to multiple requests will most likely be faster than a request and response model such as Web Services, but the implementation is far more difficult.

Now that you've looked at the terminology and a couple of connection models, take a look at the next section where the XML document in the *SOAP* message is finally discussed.

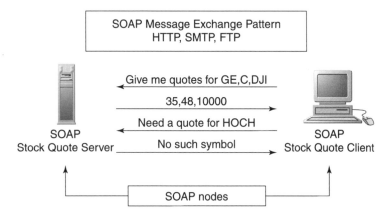

FIGURE 3.3 How two *SOAP* nodes converse over a stock quote.

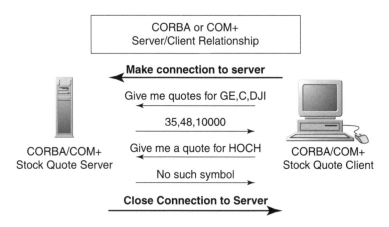

FIGURE 3.4 How a CORBA or COM+ connection can handle the same conversation over one connection.

THE *SOAP* XML DOCUMENT

SOAP uses XML to describe the data transmitted between nodes. The tags used are to describe the document, header, and body. The tags in the body are usually specific to a vendor's implementation of *SOAP* because the standard doesn't dictate the

XML appearing there. The standard does, however, dictate the information that appears in the header and the how the envelope, body, and header elements look.

Basic *SOAP* Document

Start off with a headerless *SOAP* document such as the following.

```
<?xml version='1.0' ?>
   <env:Envelope
        xmlns:env="http://www.w3.org/2001/12/SOAP-envelope">
     <env:Body>
     </env:Body>
  </env:Envelope>
```

This is a simple *SOAP* document that doesn't say anything, but note the namespace definition and the fact that there is an envelope and a body. The document is just a container for an object to get data from, process, and then put back into the document to transmit it back to the client. Let's consider the request for stock quote information again.

```
<?xml version='1.0' ?>
   <env:Envelope
    xmlns:env="http://www.w3.org/2001/12/SOAP-envelope">
     <env:Body>
     <stockquote:symbolist
          xmlns:stockquote="http://advocatemedia.com/stocks">
       <stockquote:symbol>C</stockquote:symbol>
       <stockquote:symbol>GE</stockquote:symbol>
       <stockquote:symbol>DJI</stockquote:symbol>
      <stockquote:symbolist>
     </env:Body>
  </env:Envelope>
```

Now we have information specific to the object that returns stock quotes. Note that the stockquote tags are not defined by the *SOAP* standard; rather, they are defined by the developer responsible for the software in which the Stock Quote Web Service was created.

SOAP Header and an Intermediary

The header in a *SOAP* document is optional, but when it is present it can determine the next node of a *SOAP* request when sent through an intermediary. Back in Fig-

ure 3.2, a *SOAP* intermediary looks at the contents of the XML to determine where
to route the file next.

Remember that a SOAP *intermediary can act or modify the data in the document
and then pass it to the next node, or just simply act as a router and pass it to the
next appropriate node.*

For an intermediary to pass the information along, the document must contain
the information to send it to the next node. The following *SOAP* example possesses
this information within the env:Header element.

The header defines the namespace am along with providing the actor element,
which tells the node that it should route the information somewhere else after it's
done processing it.

Then within the header there is information for the node that allows the intermediary to act on the data. In this case, a customer Id and a request Id are present
to help the initial node do some processing of the data before passing it on. It may,
for example, check the person's customer Id to make sure they are still an active
customer. The env:mustUnderstand attribute means that the node must act on the
data in some matter and return an exception if the data is not processed.

```xml
<?xml version='1.0' ?>
    <env:Envelope
        xmlns:env="http://www.w3.org/2001/12/SOAP-envelope">
      <env:Header>
        <am:customer
            xmlns:route="http://advocatemedia.com/authenticate"
            env:actor="http://www.w3.org/2001/12/SOAP-envelope/
                actor/next"
            env:mustUnderstand="true">
          <am:custId>4557799</am:custId>
          <am:requestId>12asd-34ccd-23cuden</am:requestId>
        </am:customer>
      </env:Header>

      <env:Body>
        <stockquote:symbolist
          xmlns:stockquote="http://advocatemedia.com/stocks">
          <stockquote:symbol>C</stockquote:symbol>
          <stockquote:symbol>GE</stockquote:symbol>
          <stockquote:symbol>DJI</stockquote:symbol>
        <stockquote:symbolist>
```

```
    </env:Body>
   </env:Envelope>
```

At this point, the *SOAP* document does not contain information needed for it to bind with a transport protocol.

SOAP Data Types and Structures

Along with the header, body, and envelope, the *SOAP* standard allows for the representation of certain data types and structures.

Primitive Types

The *SOAP* standard does not create new data types for variables, but, rather, uses the data types defined in the XML Schema standard. This allows *SOAP* to represent data in a standard way. Table 3.3 shows the different standard types that a *SOAP* document may represent.

TABLE 3.3   The Different Primitive Types Available from XML Schema in the *SOAP* Standard[1]

Schema Primitive Type	*Description*	*Example*
xsd:int	signed integer value	-9 or 9
xsd:boolean	boolean whose value is either 1 or 0	1 or 0
xsd:string	string of characters	Rocky Mountains
xsd:float or xsd:double	signed floating point number (+,-)	-9.1 or 9.1
xsd:timeInstant	date/time	1969-05-07-08:15
SOAP-ENC:base64	base64-encoded information used for passing binary data within *SOAP* documents	SW89ljhhibdOl111QWgdGE

[1]Adapted from Dave Winer and Jake Savin, A Busy Developer's Guide to SOAP 1.1, *http://www.SOAPware.org/bdg* (2001).

Structs

A struct is a data structure that you can think of like a container. It is a way of storing several, perhaps vastly different, values in one neat package. In fact, you saw a struct in one of the previous stockquote examples. Here is the snippet that represents a struct.

```
<stockquote:symbolist
  xmlns:stockquote="http://advocatemedia.com/stocks">
  <stockquote:symbol>C</stockquote:symbol>
  <stockquote:symbol>GE</stockquote:symbol>
  <stockquote:symbol>DJI</stockquote:symbol>
<stockquote:symbolist>
```

Modifying this snippet to use the schema's primitive types looks like the following example.

(Note that, in the header of the XML document the appropriate namespaces need to be included to use these types. Examples later in the chapter demonstrate this.)

```
<stockquote:symbolist
  xmlns:stockquote="http://advocatemedia.com/stocks">
  <stockquote:symbol
    xsi:type="string">C</stockquote:symbol>
  <stockquote:symbol
    xsi:type="string">GE</stockquote:symbol>
  <stockquote:symbol
    xsi:type="string">DJI</stockquote:symbol>
<stockquote:symbolist>
```

The information in a struct doesn't have to be so similar. For example, the XML may contain information that describes the author, such as the following.

```
<CRM:AuthorInfo xmlns:CRM="http://www.charlesriver.com/authorinfo">
    <CRM:FirstName xsi:type="string">Brian</CRM:FirstName>
    <CRM:LastName
     xsi:type="string">Hochgurtel</CRM:LastName>
    <CRM:PhoneNumber
     xsi:type="int">3035551212</CRM:PhoneNumber>
    <CRM:BookTitle
     xsi:type="string">Cross Platform Web Services</CRM:BookTitle>
  </CRM:AuthorInfo>
```

Now moving all the information together may allow an application to handle the data more efficiently.

Arrays

The *SOAP* standard also supports the use of arrays. An array is similar to a struct but normally only stores data of the same type, like the following example.

```
<SymbolList
SOAP-ENC:arrayType="xsd:string[3]">
 <symbol>C</symbol>
    <symbol>GE</symbol>
    <symbol>DJI</symbol>
</SymbolList>
```

This is an alternate, and perhaps more convenient, way to represent data for the Stock Quote Example. Instead of having to define the type for each entry, the array definition allows you to group related data together so you don't have to specify the type each time.

However, it is possible to group unlike values in an array as the following example illustrates.

```
<AuthorInfo SOAP-ENC:arrayType="xsd:ur-type[4]" >
    <FirstName xsi:type="string">Brian</FirstName>
    <LastName
     xsi:type="string">Hochgurtel</LastName>
    <PhoneNumber
     xsi:type="int">3035551212</PhoneNumber>
    <BookTitle xsi:type="string">
        Cross Platform Web Services
    </BookTitle>
<AuthorInfo>
```

The definition for the array type, xsd:ur-type[4], indicates that there are four elements in the array of various types. This is much different than the struct type introduced earlier in the chapter.

For one final look at arrays, consider the following example that shows the entire Stock Quote Example using an array. Note that to use the schema type string in the document, the namespaces for schemas must be included in the header.

```
<?xml version='1.0' ?>
<env:Envelope
  xmlns:env="http://www.w3.org/2001/12/SOAP-envelope"
  xmlns:xsd="http://www.w3.org/1999/XMLSchema"
  xmlns:xsi="http://www.w3.org/1999/XMLSchema-instance"
```

```
     xmlns:SOAP-ENC="http://schemas.xmlSOAP.org/SOAP/encoding/"
     xmlns:stockquote="http://advocatemedia.com/examples">
     <env:Body>

        <SOAP-ENC:Array SOAP-ENC:arrayType="xsd:string[3]">
          <stockquote:symbol>C</stockquote:symbol>
          <stockquote:symbol>GE</stockquote:symbol>
          <stockquote:symbol>DJI</stockquote:symbol>
        </SOAP-ENC:Array>
     </env:Body>
   </env:Envelope>
```

Using the array, the format is slightly cleaner and easier to read, but if you look at the original Stock Quote Example that uses the struct, there really isn't much difference.

SOAP DOCUMENTS AND TRANSPORT BINDINGS

In addition to the XML in a *SOAP* request, there is also a header outside of the XML that is specific for the protocol being used, such as HTTP. The information in this header contains the response code, the version of the protocol being used, the content type of the message, and perhaps other vendor-specific information. The next section shows the type of header needed for an HTTP request.

HTTP Request

The following example shows the Stock Quote *SOAP* Request with its HTTP header. This information tells the server receiving the request where to send the request based on the information here.

The first line states that the information is being posted to the Stock Quotes Web Service using HTTP Version 1.0. The host information on the next line tells the request where on the *World Wide Web* (WWW) the service exists. The Content-Type definition shows what type of information is in the request. In this case, it is XML. Content-length tells how many characters exist in the request. SOAP action contains the namespace for this particular Web Service (*http://www.advocatemedia. com/webservices/*) and the name of the particular method responsible for processing the data (getquote).

```
POST /stockquotes HTTP/1.1
Host: www.advocatemedia.com:80
Content-Type: text/xml; charset=utf-8
Content-Length: 482
SOAPAction: "http://www.advocatemedia.com/webservices/getquote"

<?xml version='1.0' ?>
<env:Envelope
     xmlns:env="http://www.w3.org/2001/12/SOAP-envelope"
     xmlns:xsd="http://www.w3.org/1999/XMLSchema"
     xmlns:xsi="http://www.w3.org/1999/XMLSchema-instance"
     xmlns:SOAP-ENC="http://schemas.xmlSOAP.org/SOAP/encoding/"
     xmlns:stockquote="http://advocatemedia.com/examples">
<env:Body>
  <SOAP-ENC:Array SOAP-ENC:arrayType="xsd:string[3]">
    <stockquote:symbol>C</stockquote:symbol>
    <stockquote:symbol>GE</stockquote:symbol>
    <stockquote:symbol>DJI</stockquote:symbol>
  </SOAP-ENC:Array>
 </env:Body>
</env:Envelope>
```

The header is what makes this example go from an XML document to an actual
SOAP request.

HTTP Response

The following code is a possible response to the Stock Quote Example. The header
contains information similar to the request, but not as much information is needed
in the response because the connection still knows where the client resides.

The first part of the header indicates that the response comes back via HTTP
Version 1.1 and that the status is 200 (which means complete) and OK meaning the
data processed correctly. The second line shows that now the request and response
are complete the connection is closed. The final two lines are just like the request
where Content-Length indicates the number of characters in the message and
Content-Type indicates that the response contains XML.

```
HTTP/1.1 200 OK
Connection: close
Content-Length: 659
Content-Type: text/xml; charset=utf-8
```

```
<?xml version='1.0' ?>
<env:Envelope
  xmlns:env="http://www.w3.org/2001/12/SOAP-envelope"
  xmlns:xsd="http://www.w3.org/1999/XMLSchema"
  xmlns:xsi="http://www.w3.org/1999/XMLSchema-instance"
  xmlns:SOAP-ENC="http://schemas.xmlSOAP.org/SOAP/encoding/"
  xmlns:stockquote="http://advocatemedia.com/examples">
  <env:Body>
      <SOAP-ENC:Array SOAP-ENC:arrayType="xsd:int[3]">
        <stockquote:price
         stockquote:symbol="C">53.21</stockquote:price>
        <stockquote:price
         stockquote:symbol="GE">48.00</stockquote:price>
        <stockquote:price
          stockquote:symbol="DJI">9500</stockquote:price>
      </SOAP-ENC:Array>
  </env:Body>
</env:Envelope>
```

Notice that the response uses a *SOAP* array to return data.

An Example of a *SOAP* Error

If there's an error, like passing a nonexistent stock symbol, the *SOAP* standard needs to have a mechanism to deal with that. The following example shows what happens with both the XML document and the server header.

The HTTP header indicates that an error occurred during the request. Not only are there the words "Server Error" but also code 500 indicates that there was an error. Within the XML, there are two tags to indicate what happened. The faultcode element contains information that the software can parse and understand what happened. The faultstring element tells the programmer what happened. The following, for example, would be the text that would show up in some sort of pop-up error window in an application.

```
HTTP/1.1 500 Server Error
   @dis:Connection: close
   Content-Length: 511
   Content-Type: text/xml; charset=utf-8
   <env:Envelope
     xmlns:env="http://www.w3.org/2001/12/SOAP-envelope"
     xmlns:xsd="http://www.w3.org/1999/XMLSchema"
```

```
        xmlns:xsi="http://www.w3.org/1999/XMLSchema-instance"
        xmlns:SOAP-ENC="http://schemas.xmlSOAP.org/SOAP/encoding/">

<env:Body>
  <env:Fault>
    <env:faultcode>error271</env:faultcode>
    <env:faultstring>No such ticker symbol</env:faultstring>
  </env:Fault>
</env:Body>
```

This example utilizes HTTP as the transport, but the `faultcode` and `fault-string` can be matched with any protocol.

SOAP and SMTP

The *SOAP* standard allows the XML in the message to bind with protocols other than HTTP. This is especially useful if you work on a project that deals with countries that do not have good connections to the Internet. Many times users in these countries, such as many on the African continent, will have the connections time out through HTTP, but with e-mail and *Standard Mail Transport Protocol* (SMTP) the message can take its time getting to the receiver because e-mail gets broken up into several pieces as it works its way through the Internet. HTTP, on the other hand, is looking for an immediate response within a fairly immediate timespan.

To work with SMTP, a *SOAP* document needs to replace the HTTP header with information needed for e-mail. Consider our Stock Quote Request again with a SMTP header.

Rather than having information needed for HTTP, such as `Post` and the URL of the service, the SMTP *SOAP* header contains an e-mail address, a subject, and a date. A *SOAP* request may also contain unique message `Ids`. This is just like sending an e-mail to a person except that software generated the e-mail to send to the receiver. In addition, instead of a text message, the application sends the *SOAP* document that contains the XML needed for the Stock Quote Web Service.

```
From: brian@advocatemedia.com
To: stockquotes@advocatemedia.com
Subject: stock quotes
Date: 18 JUN 2002 18:00:00 MDT
<?xml version='1.0' ?>
<env:Envelope
  xmlns:env="http://www.w3.org/2001/12/SOAP-envelope"
```

```
xmlns:xsd="http://www.w3.org/1999/XMLSchema"
xmlns:xsi="http://www.w3.org/1999/XMLSchema-instance"
xmlns:SOAP-ENC="http://schemas.xmlSOAP.org/SOAP/encoding/"
xmlns:stockquote="http://advocatemedia.com/examples">
<env:Body>
    <SOAP-ENC:Array SOAP-ENC:arrayType="xsd:int[3]">
      <stockquote:price
       stockquote:symbol="C">53.21</stockquote:symbol>
      <stockquote:price
       stockquote:symbol="GE">48.00</stockquote:symbol>
      <stockquote:price
        stockquote:symbol="DJI">9500</stockquote:symbol>
    </SOAP-ENC:Array>
  </env:Body>
</env:Envelope>
```

The response to this request, again, contains the header for SMTP with the exact same response in the *SOAP* document as in previous examples. The subject should change to indicate that some sort of process actually occurred.

The request and response sent via e-mail is read and created by applications of the SOAP nodes. You wouldn't want the users of your application decoding some XML in their e-mail inbox in exchange for the data they wanted.

The message is simply returned to the sender with the appropriate XML in the body of the e-mail message.

```
From: stockquotes@advocatemedia.com
To: brian@advocatemedia.com
Subject: your requested stock quotes
Date: 18 JUN 2002 18:00:00 MDT
<?xml version='1.0' ?>
<env:Envelope
  xmlns:env="http://www.w3.org/2001/12/SOAP-envelope"
  xmlns:xsd="http://www.w3.org/1999/XMLSchema"
  xmlns:xsi="http://www.w3.org/1999/XMLSchema-instance"
  xmlns:SOAP-ENC="http://schemas.xmlSOAP.org/SOAP/encoding/"
  xmlns:stockquote="http://advocatemedia.com/examples">
  <env:Body>
      <SOAP-ENC:Array SOAP-ENC:arrayType="xsd:int[3]">
        <stockquote:price
```

```
        stockquote:symbol="C">53.21</stockquote:symbol>
       <stockquote:price
        stockquote:symbol="GE">48.00</stockquote:symbol>
       <stockquote:price
         stockquote:symbol="DJI">9500</stockquote:symbol>
      </SOAP-ENC:Array>
    </env:Body>
   </env:Envelope>
```

Having a choice between protocols gives a developer a great deal of flexibility of how to transmit data between nodes.

CONCLUSION

In this chapter, the various aspects of the *SOAP* standard were covered. The important part to realize is that the *SOAP* standard covers how the XML in the document needs to appear, how the document is transmitted, the vocabulary of the document exchange, and with many other requirements. Thus, when you see the term *SOAP*, you should now realize that it's more than just an XML document.

4 WSDL

W SDL stands for *Web Services Description Language*, and it is a method of describing a Web Service using XML. For Web Services to work cross platform, they need to have a standard way to transmit and describe what happens within the Web Service. WSDL provides the means to describe what messages and variables exist within the Web Service whereas Universal Discovery, Description, and Integration (UDDI) is a standard way of publishing a Web Service to your desired audience. WSDL is part of the UDDI standard; Chapter 5 covers this in greater detail.

Having a standard way of describing a remote object is not unique to Web Services. CORBA has the *Interface Description Language* (IDL) that describes the CORBA object to the client, and Microsoft has the *Microsoft Interface Description Language* (MIDL) for COM+. IDL and, to a lesser extent, MIDL conform to the standard set by the Object Management Group (OMG). The OMG manages certain standards that the industry uses, just like the W3C, where particular aspects of a particular protocol are maintained by a committee and the results are published on the Web.

Figure 4.1 shows how a Web Service client uses a WSDL file. Even though this example is specific to Web Services, this interaction is similar for all remote object technologies.

Before any Web Service call is made, the user creating the client uses a tool to grab the WSDL file off the server; this is shown in the first transaction of Figure 4.1. Many Web Service technologies have a standard way of revealing their WSDL file so it isn't difficult for a user to find.

Once the user has the WSDL file, some technologies, such as Microsoft's *.NET*, allow the user to then create a proxy to the Web Service. This proxy then acts as an interface to the Web Service, allowing the client-side code to easily access the Web Service. In the case of Figure 4.1, the WSDL file reveals to the client that the Stock Quote server has three methods available: `Get Quote`, `Active Account`, and `Get Research`. The client now has an interface to each of these methods and is able to use them.

This chapter breaks down each individual section of the WSDL file that describes the `GetStockQuote` Web Service used in examples in the previous chapters.

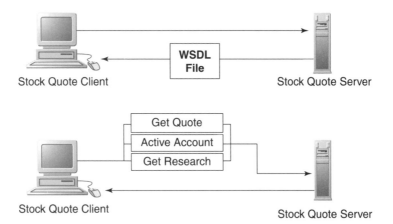

FIGURE 4.1 How a client asks for a WSDL file and uses it to create an interface to the methods from the Web Service.

REVISITING THE STOCK QUOTE WEB SERVICE

Before you begin to learn WSDL, it is necessary to revisit the GetStockQuote Web Service that was discussed in the last chapter because the WSDL example in this chapter builds on this document.

In the last chapter, the Web Service received several requests for quotes and then returned several prices. For the sake of simplifying the WSDL code, the *SOAP* code for the example in this chapter simply receives one stock symbol and then transmits one price.

In the following code, you'll see that the single symbol "C" (for Citicorp) resides in the code.

```
POST /stockquotes HTTP/1.1
 Host: www.advocatemedia.com:80
 Content-Type: text/xml; charset=utf-8
 Content-Length: 482
 SOAPAction:    "http://www.advocatemedia.com/webservices/getquote"

 <?xml version='1.0' ?>
 <env:Envelope
   xmlns:env="http://www.w3.org/2001/12/soap-envelope"
   xmlns:xsd="http://www.w3.org/2001/XMLSchema"
```

```
      xmlns:xsi="http://www.w3.org/2001/XMLSchema-instance"
      xmlns:SOAP-ENC="http://schemas.xmlsoap.org/soap/encoding/">
   <env:Body>
      <StockQuoteRequest xmlns="http://advocatemedia.com/Examples">
        <symbol xsi:type="string">
           C
        </symbol>
      </StockQuoteRequest>
   </env:Body>
   </env:Envelope>
```

The response sends one single price back to the client. In the following example, you'll see that the service sends the price of the stock back to the client in the form of float.

```
HTTP/1.1 200 OK
Connection: close
Content-Length: 659
Content-Type: text/xml; charset=utf-8

<?xml version='1.0' ?>
<env:Envelope
   xmlns:env="http://www.w3.org/2001/12/soap-envelope"
   xmlns:xsd="http://www.w3.org/2001/XMLSchema"
   xmlns:xsi="http://www.w3.org/2001/XMLSchema-instance"
   xmlns:SOAP-ENC="http://schemas.xmlsoap.org/soap/encoding/">
   <env:Body>
     <StockQuoteResponse xmlns="http://advocatemedia.com/Examples"

     <price xsi:type="float">
        53.21
     </price>
     </StockQuoteResponse>
   </env:Body>
   </env:Envelope>
```

Thus, overall the structure of the Web Service is the same. It's just simplified for the examples in this chapter.

STRUCTURE OF THE WSDL DOCUMENT

The WSDL document describes each piece of the Web Service from each of the elements found in the XML (besides the standard *SOAP* elements) to the transport and the name.

The main six parts or elements of the WSDL document include: `definitions`, `types`, `message`, `portType`, `binding`, and `service`. Table 4.1 summarizes what each section describes.

TABLE 4.1 The Parent Elements That Make up an XML Document

Element	Purpose
definitions	The root element of the WSDL document. Defines many of the namespaces used for a particular description.
types	Defines the elements and primitive types found in the XML of the *SOAP* request and response.
message	Names the request and response messages.
portType	Ties a particular request and response message to a particular service.
binding	Indicates the type of transport (i.e., HTTP) used by the service. Also describes the contents of the message such as `literal`, meaning to take the XML at face value, or `image/gif`.
service	Actually names the services and provides an element to document what the service actually accomplishes.

Each of the sections is covered in greater detail later in the chapter. To help illustrate the structure of the WSDL document, Figure 4.2 uses a diagram.

You'll notice in Figure 4.2 that several of the elements actually have child elements as well. For example, the `types` element has an element that describes the elements found in the XML; it may contain several elements for a Web Service that contains many methods.

Now that you have been briefly introduced to each section of a WSDL document, the following sections go into greater detail.

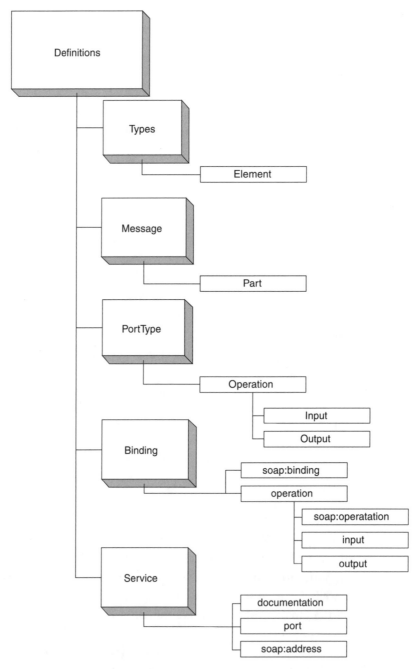

FIGURE 4.2   Illustrates the structure of the WSDL document along with any child elements of each major section.

definitions

`definitions` is the root element of the WSDL document. The beginning of the document may look like the following.

```
<definitions name="GetStockQuote></definitions>
```

This only defines the name of the Web Service. To be useful, however, the `definitions` tag will need to define several different namespaces to support the different primitive types and the types created by the `GetStockQuote` Web Service. Once the namespaces are added, the `definitions` element now looks like the following.

```
<definitions name="GetStockQuote
    targetNamespace="http://advocatemedia.com/GetStockQuote.wsdl"
    xmlns:myns = "http://advocatemedia.com/GetStockQuote.wsdl"
    xmlns:myXsd = "http://advocatemedia.com/GetStockQuote.xsd"
    xmlns:soap = "http://schemas.xmlsoap.org/wsdl/soap"
    xmlns="http://schemas.xmlsoap.org/wsdl/">
</definitions>
```

Table 4.2 describes what each namespace definition does for the WSDL document.

TABLE 4.2   Detailed Information about What Each Namespace Definition Does for the WSDL Document

Namespace Definition	Placeholder	Description
`targetNamespace`	*http://advocatemedia.com/ GetStockQuote.wsdl*	Defines the namespace for this document.
`Myns`	*http://advocatemedia.com/ GetStockQuote.wsdl*	A more precise definition for this document.
`MyXSD`	*http://advocatemedia.com/ GetStockQuote.xsd*	Namespace for the schema types defined here.
`xmlns:soap`	*http://schemas.xmlsoap.org/ wsdl/soap*	Namespace for the *SOAP* elements used in the document.
`xmlns`	*http://schemas.xmlsoap.org/ wsdl/*	Sets the default namespace for *SOAP* elements.

Remember that a namespace is just a unique placeholder for a set of elements in an XML document. The URL is just a unique value and is not necessarily a URL to which you may point your browser.

types

The types element defines the different elements used within the Web Service. The elements defined here are for the variables in our Stock Quote Web Service. Just like with *SOAP*, WSDL uses the types from the schema standard so that standard types are used.

The types element defines a schema in the middle of our WSDL document. By doing this, a WSDL document is able to use the types defined in the schema standard rather than having to create its own.

Take a look at the following example and notice that several elements are defined.

```
<types>
    <schema
        targetNamespace="http://advocatemedia.com/GetStockQuote.xsd"
        xmlns="http://www.w3.org/2000/10/XMLSchema">
        <element name="StockQuoteRequest">
            <complexType>
                <all>
                    <element name="symbol" type="string"/>
                </all>
            </complexType>
        </element>
        <element name="StockQuoteResponse">
            <complexType>
                <all>
                    <element name="price" type="float"/>
                </all>
            </complexType>
        </element>
    </schema>
</types>
```

StockQuoteRequest and StockQuoteResponse are the parent elements of the request and response documents. The elements that actually contain values (i.e., symbol and price), are the children. The types for these child elements also are defined here. symbol is defined as a string and price is defined as a float.

message

Now the WSDL document needs to describe and name both the request and response message. This comes after the type definition and gives a path back to the types in the message. The message elements look like the following.

```
<message name="GetStockQuoteRequest">
    <part name="body" element="myXSD:StockQuoteRequest"/>
</message>

<message name="GetStockQuoteResponse">
    <part name="body" element="myXSD:StockQuoteResponse"/>
</message>
```

Notice that the element definitions have the names of the parent elements in the *SOAP* document. The namespace myXSD is used as a prefix so that the application using the WSDL document knows to find the definitions for these elements in the types portion of the document.

This gives the request and response messages a name so the applications or Web pages using this service know the name of the message to send and expect back when using a particular service. Note that this is protocol independent because there is no mention of HTTP or SMTP.

The value of the name can be anything, but you should select something that is meaningful to you.

NOTE

portType

To use the two messages defined in the previous section with the message element, you must define them as the request and response for a particular service.

This is done with the portType command, as shown in the following example.

```
<portType name="GetStockQuotePort">
    <operation name="GetStockQuote">
        <input message="myns:GetStockQuoteRequest"/>
        <output message="myns:GetStockQuoteResponse"/>
    </operation>
</portType>
```

GetStockQuote is now considered the name of the operation. The operation is the request for the price of a particular stock and the response is in the form of a price for that stock. The input and output messages just combine the two definitions used earlier so that the client knows that a particular request and response message belongs to a particular method in a Web Service.

If this were a more complex Web Service, there would be several portType, message, and other elements. The example we're using here is very simple so the WSDL isn't too complex.

binding

The binding in a WSDL document indicates the type of transport that a Web Service uses. For example, if an XML document transmits its contents in the body, then the WSDL document needs to define that. If the document transmits its contents in Base64 encoding, then that would need to be defined here as well.

The binding name can be anything you wish. In this case, the name is GetStockQuoteBindingName.

Tools generating WSDL will use their own algorithms for naming the different sections.

The next element is soap:binding. It defines the style of the binding and the transport. There are two styles to choose from: RPC and document. RPC indicates that a message contains parameters and that there are return values. When the style attribute contains document, the request and response are passing XML documents within the body of the *SOAP* message.

The namespace in the transport indicates which protocol is used for the Web Service. In this case, we use HTTP by looking at the namespace. This is the required namespace for the HTTP transport. If using another transport such as SMTP, you can specify your own namespace, but the name must contain the transport like this: *http://advocatemedia.com/smtp/*. It is interesting to note that a WSDL document must specify the transport, but it cannot specify the address of the service according to the standard.

The next part of the WSDL document, where the input and output elements reside, defines the contents of the request and response messages. In this case, the example uses the contents of the document without any encoding. If either message contained any encoding such as Base64 or image/gif, it would be indicated here.

```
<binding name="GetStockQuoteBindingName" type="GetStockQuotePort ">
    <soap:binding
        style="rpc"
        transport=" http://schemas.xmlsoap.org/soap/http"/>
    <operation name="GetStockQuote">
        <soap:operation
            soapAction="http://advocatemedia.com/GetStockQuote"/>
        <input>
            <soap:body use="literal"/>
        </input>
        <output>
            <soap:body use="literal"/>
        </output>
    </operation>
</binding>
```

The only part of the Web Service left to define in the WSDL is the actual service.

Defining the `service`

Now that the transport, actions, content, style of message, and many other things have been defined, it is time to actually define the service. The service element, as shown in the following code example, actually defines the name of the service along with documentation and the location of the service. The following is the code.

```
<service name="AdvocateMediaGetStockQuotes">
    <documentation>Simple Web Service to
                    Retrieve a stock quote</documentation>
    <port name="GetStockQuotePort"
        binding="myns:GetStockQuoteBindingName">
      <soap:address
          location="http://advocatemedia.com/GetStockQuote"/>
    </port>
</service>
```

The documentation element gives a developer the opportunity to provide some added information about what the Web Service accomplishes. The soap:address names the Web Service as a whole.

THE COMPLETE WSDL FILE

Now that you have examined each section of a WSDL document, you need to look
at it as a whole. Remember that WSDL is not necessarily for a person to read; it's
more for an application to read in order to use the Web Service. Therefore, the
XML might look pretty ugly in the complete document.

```xml
<definitions name="GetStockQuote
        targetNamespace="http://advocatemedia.com/GetStockQuote.wsdl"
        xmlns:myns = "http://advocatemedia.com/GetStockQuote.wsdl"
        xmlns:myXsd = "http://advocatemedia.com/GetStockQuote.xsd"
        xmlns:soap = "http://schemas.xmlsoap.org/wsdl/soap"
        xmlns="http://schemas.xmlsoap.org">
    <types>
    <schema
        targetNamespace="http://advocatemedia.com/GetStockQuote.xsd"
        xmlns="http://www.w3.org/2000/10/XMLSchema">
        <element name="StockQuoteRequest">
            <complexType>
              <all>
                <element name="symbol" type="string"/>
              </all>
            </complexType>
        </element>
        <element name="StockQuoteResponse">
            <complexType>
              <all>
                <element name="price" type="float"/>
              </all>
            </complexType>
        </element>
      </schema>
    </types>
    <message name="GetStockQuote">
        <part name="body" element="myXSD:StockQuoteRequest"/>
    </message>
    <message name="GetStockQuoteResponse">
        <part name="body" element="myXSD:StockQuoteResponse"/>
    </message>
    <portType name="GetStockQuotePort">
        <operation name="GetStockQuote">
            <input message="myns:GetStockQuoteRequest"/>
```

```
            <output message="myns:GetStockQuoteResponse"/>
        </operation>
    </portType>
    <binding name="GetStockQuoteBindingName"
             type="StockQuoteBinding">
    <soap:binding
         style="rpc"
         transport=" http://schemas.xmlsoap.org/soap/http"/>
        <operation name="GetStockQuote">
            <soap:operation
                soapAction="http://advocatemedia.com/GetStockQuote"/>
            <input>
                <soap:body use="literal"/>
            </input>
            <output>
                <soap:body use="literal"/>
            </output>
        </operation>
    </binding>
    <service name="AdvocateMediaGetStockQuotes">
      <documentation>Simple Web Service to
                   Retrieve a stock quote</documentation>
      <port name="GetStockQuotePort"
            binding="myns:GetStockQuoteBindingName">
        <soap:address
            location="http://advocatemedia.com/GetStockQuote"/>
      </port>
    </service>
</definitions>
```

Using `import`

There is an alternate and perhaps easier way of writing the WSDL file. This involves defining all the types in an *XML Schema Reduced* (XSD) file while putting all the other definitions relevant to the Web Service in the WSDL file. This way, the schema element and all of its children are in a separate file. If you are using the same elements in different Web Services, you can easily move the schema definitions from application to application. For example, there may be several stock-related Web Services that use the same types and variables. This way one XSD file could support all the different services.

Here is the GetStockQuote.xsd example.

```
<schema
    targetNamespace="http://advocatemedia.com/GetStockQuote.xsd"
    xmlns="http://www.w3.org/2000/10/XMLSchema">
    <element name="StockQuoteRequest">
      <complexType>
        <all>
          <element name="symbol" type="string"/>
        </all>
      </complexType>
    </element>
    <element name="StockQuoteResponse">
      <complexType>
        <all>
          <element name="price" type="float"/>
        </all>
      </complexType>
    </element>
</schema>
```

From within the WSDL file, the `import` element is used to bring in the element definitions in GetStockQuote.xsd, as shown in the following code.

```
<definitions name="GetStockQuote
    targetNamespace="http://advocatemedia.com/GetStockQuote.wsdl"
    xmlns:myns = "http://advocatemedia.com/GetStockQuote.wsdl"
    xmlns:myXsd = "http://advocatemedia.com/GetStockQuote.xsd"
    xmlns:soap = "http://schemas.xmlsoap.org/wsdl/soap"
    xmlns="http://schemas.xmlsoap.org">
    <import namespace="http://advocatemedia.com/GetStocks/schemas"
           location ="http://advocatemedia.com/GetStocks/quote.xsd">
    <message name="GetStockQuote">
      <part name="body" element="myXSD:StockQuoteRequest"/>
    </message>
    <message name="GetStockQuoteResponse">
      <part name="body" element="myXSD:StockQuoteResponse"/>
    </message>
    <portType name="GetStockQuotePort">
        <operation name="GetStockQuote">
            <input message="myns:GetStockQuoteRequest"/>
            <output message="myns:GetStockQuoteResponse"/>
        </operation>
    </portType>
    <binding name="GetStockQuoteBindingName"
```

```
                    type="StockQuoteBinding">
         <soap:binding
             style="rpc"
             transport=" http://schemas.xmlsoap.org/soap/http"/>
           <operation name="GetStockQuote">
              <soap:operation
                 soapAction="http://advocatemedia.com/GetStockQuote"/>
              <input>
                 <soap:body use="literal"/>
              </input>
              <output>
                 <soap:body use="literal"/>
              </output>
           </operation>
        </binding>
        <service name="AdvocateMediaGetStockQuotes">
          <documentation>Simple Web Service to
                        Retrieve a stock quote</documentation>
          <port name="GetStockQuotePort"
               binding="myns:GetStockQuoteBindingName">
            <soap:address
                 location="http://advocatemedia.com/GetStockQuote"/>
          </port>
        </service>
     </definitions>
```

Notice that only the difference between this and the original complete WSDL example is the import element.

CONCLUSION

WSDL provides the developer with the ability to describe a Web Service in great detail, starting with defining each element that appears in the body of the *SOAP* document, the names of the request and response message, the type of transport, the encoding used within the message, and the name of the service. The resulting XML looks complex because it's really not meant for human consumption. The WSDL is meant for use by an application during the discovery phase of developing a new Web Services client.

5 ∷ UDDI

In This Chapter

- Before UDDI and *SOAP*
- Finding a Stock Quote Web Service
- UDDI Involves Teamwork
- Internal UDDI Web Sites
- A UDDI Case Study
- Don't Forget the XML

U DDI stands for *Universal Discovery, Description, and Integration*. UDDI is a repository of Web Services and companies that provides Web Services to the public and other companies. UDDI also contains business information for corporations that work in this arena. The repository concept evolved from the concept of *Business to Business* (B2B) exchanges working furiously to collaborate across the Web. Prior to the UDDI standard, there was not a way for these exchanges to easily find each other, much less the functionality they needed to share. By having a standard place in which a corporation may search for partners, the collaboration process moves forward at a quicker pace.

Several major vendors host the UDDI repositories, such as Microsoft and IBM, as a service to the Web Service community. There is currently no charge to use any part of the implementation. Each vendor that hosts a repository ends up replicating with the other partners so that they should contain the same information. Therefore, when you use the repository it shouldn't matter which vendor's implementation you consider.

You contact the repository in one of two ways: either through a Web browser or through *SOAP* requests and responses by using a particular API. This gives you great flexibility because you can either search yourself or create tools to automatically search the repository on a regular basis, looking for functionality you might need.

The UDDI standard confuses many because it encompasses many things, much like the *SOAP* standard. *SOAP* not only describes the contents of a Web Service but also describes how the messages are transmitted and much more. UDDI defines how a repository operates and how the Web Services and corporations are described with XML along with the API for contacting the repository with an application.

The repository of information contains three different sets of data. The first is general contact information, which includes street address and perhaps an individual responsible for fielding inquiries. Searching for the type of service, such as stock quotes, is also available along with information about the type of industry a particular company resides in. By thinking of UDDI as a combination of the white and yellow pages for Web Services and businesses, you begin to get a clearer picture of how to take advantage of UDDI.

Links to WSDL files are often found at a UDDI site, but they do not share any formal relationship. UDDI tells you where the Web Service resides and who sponsors it. On the other hand, WSDL describes the Web Service, including which methods are available. When trying to discover Web Services, you often find that a UDDI site leads you to a WSDL file—that's why the misconception that they are related exists.

If you need to be compatible with Microsoft's .NET environment, look in the repository for Web Services that have a URL for the WSDL file. .NET Web Services need this to create a proxy dll (or a Web reference in Visual Studio.NET*) to use the methods in that Web Service. Chapter 6 introduces .NET Web Services and describes how to use the WSDL file appropriately.*

BEFORE UDDI AND *SOAP*

With COM and CORBA implementations, there is no standard way to find documentation or the locations of a particular remote object. It is a matter of finding the designated developer or administrator who maintains the object and gathering whatever documentation they possess. The documentation may look like the following.

```
Responsible Developer: Brian Hochgurtel
COM+ Object: GetDowJonesInfo
COM+ Server: http://advocatemedia.com/
Relative URL: /stocks/GetDowJonesInfo
Proxy location: \\install\GetDowJonesInfo.dll

Description: Returns stock quotes from Dow Jones
Industrials index. For use by internal and external
customers.
```

The style of documenting remote objects varies from corporation to corporation and can even differ between departments. These inconsistencies led to the standards committees developing UDDI because it provides a standard for describing the location and purpose of a Web Service, along with other business information. Using UDDI in conjunction with WSDL enables you to have a complete description of the Web Service's sponsoring company, the information available from the Web Service, other Web Services provided, and a description of each method in the service.

FINDING A STOCK QUOTE WEB SERVICE

Throughout this book, the stock quote Web Service is mentioned as an example. So far it has just been theoretical—an easy example to help you understand what a Web Service is meant to do. Now that you understand a little bit about UDDI, you can search for a real stock quote Web Service. Start out by going to Microsoft's UDDI Web site at *http://uddi.microsoft.com*. Figure 5.1 shows you the home page.

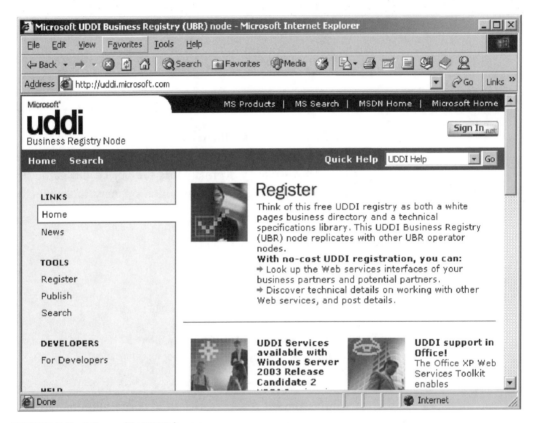

FIGURE 5.1 Microsoft's UDDI home page.

In the upper lefthand corner, there is a link for search. Click there and you'll see the search page shown in Figure 5.2.

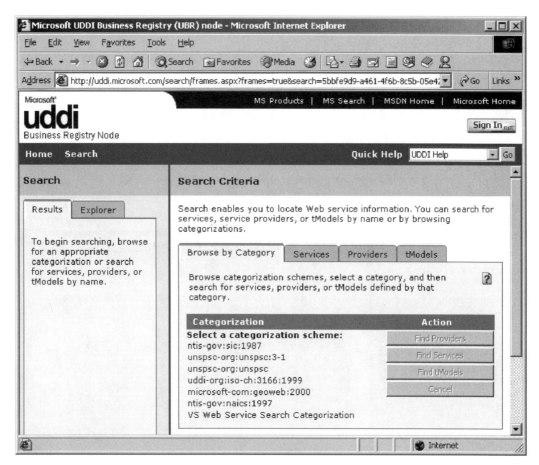

FIGURE 5.2   The search page at *http://uddi.microsoft.com*.

On the righthand side of the frame set shown in Figure 5.2, select the services tab so that you can search by the type of service. Then enter stock quote as the type of service that you're searching for.

Note that there are several ways that you can search for information on Microsoft's UDDI Web site, as you can see on the tabs on the righthand side of the site. You can browse by category, service, provider, and tModel. tModel stands for technical model. It represents a reusable idea, including the type of Web Services, which protocol a Web Service uses, or some sort of category such as providing stock values or even what size bicycle you should purchase.

The search for stock quotes under services brings back several choices for you to use (see Figure 5.3).

The results show up on the lefthand side of the Web page. Start browsing them to find a Web Service that has a WSDL file, so you can use a .NET Web Service.

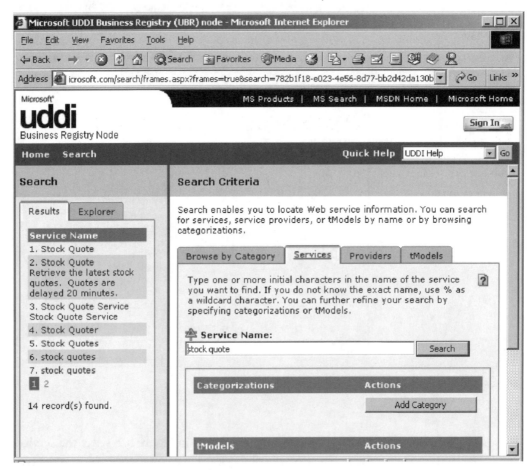

FIGURE 5.3   The results of the search for stock quote Web Services.

Any Web Service whose suffix ends with "asmx" indicates that it is an implementation of the Web Services available in Microsoft's .NET environment. By simply adding "WSDL" in the query string, the WSDL file is retrieved for you. For example, let's say that you have a Web Service at the following URL: http://localhost/Xplatform/ SomeService.asmx. *By simply changing the URL to the following, the WSDL file is revealed:* http://localhost/Xplatform/SomeService.asmx?WSDL.

By clicking around to search, you may not find the Web Service that you need. Another way to search the repository is by provider. Go back to the search page and select the tab on the lefthand side that says "providers." Enter the name "Cdyne" into the text box that asks for provider name and then click on search. Figure 5.4 shows the services that "Cdyne" makes available.

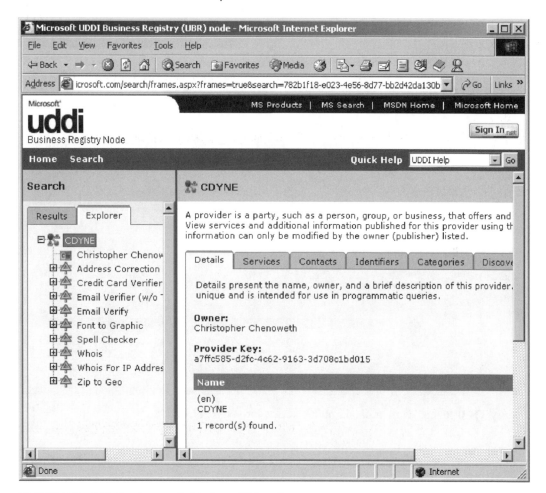

FIGURE 5.4 Web Services available from "Cdyne."

In this example, the company "Cdyne" offers a stock quote Web Service. However, instead of just a link to the service, the information on this page gives some details about the company and actual contact information. This includes a contact if there is a problem reaching the Web Service.

Clicking on the "stock quotes" link gives you the information needed to actually utilize the Web Service. Figure 5.5 displays the results of this search in *Internet Explorer*.

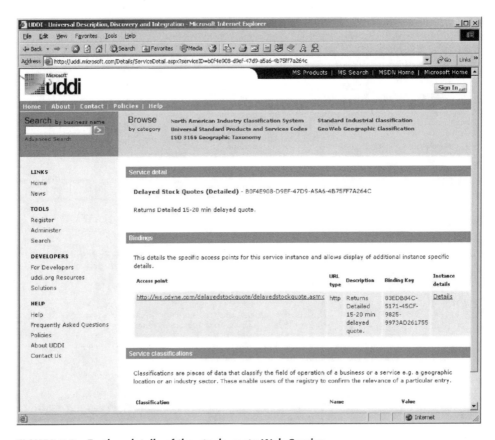

FIGURE 5.5   Further details of the stock quote Web Service.

This page on the UDDI site provides greater detail to the service; you can see that the service provides stock quotes that are 15 to 20 minutes delayed. The long alpha-numeric values in the "Binding Key" column next to the "Service Detail" are unique keys generated by the UDDI site when creating the entry. All of this information can be found by clicking through the various tabs in the righthand frame for the stock quote Web Service.

For even more information on this company, click on "Cdyne" in the lefthand pane of the frame set and then click on "discovery URLs" on the righthand frame set. Click on one of the discovery URLs and you'll find the XML for the UDDI entry for this company. Figure 5.6 displays the UDDI XML in *Internet Explorer*.

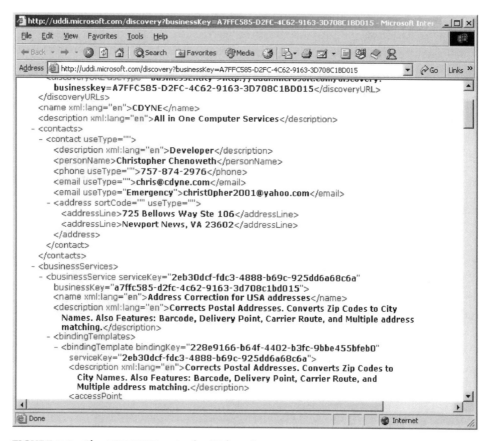

FIGURE 5.6   The UDDI XML entry for "Cdyne."

This provides some further information, but the item of main importance here is the link to the WSDL file. This is useful for determining what methods are available and for adding references or creating proxies in *Visual Studio.NET*. Note that the suffix on the Web Service is "asmx."

UDDI INVOLVES TEAMWORK

The examples in this chapter use Microsoft's UDDI site, but UDDI Web sites are supported by a group of industry leaders (including HP, IBM, Microsoft, and SAP), who formed a consortium to maintain the UDDI registry system. This system is a database of businesses and the URL of the Web Services they offer, any available WSDL files, and provided business information. Each of the vendors replicates the entries it receives back to the others. Regardless of where you enter the information about your Web Service, it should show up on the other sites as well. Figure 5.7 demonstrates the replication among the different UDDI sites.

UDDI For Discovery and Registration

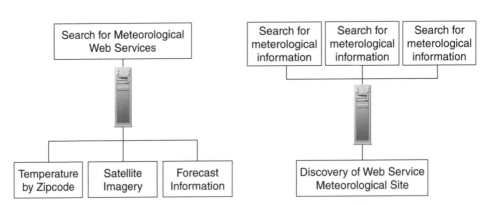

FIGURE 5.7 The replication of UDDI information amongst the support Web Sites.

The current working version of UDDI is Version 1. Table 5.1 displays the URLs of the Web sites and the vendors maintaining Web sites for this version.

TABLE 5.1 Current Implementations of UDDI Version 1

Vendor	URL
Microsoft	http://uddi.microsoft.com/
IBM	http://www-3.ibm.com/services/uddi/find

UDDI Version 2 should soon be available for production; as of this writing the UDDI Version 2 sites are in Beta. The main difference between the two different versions seems to be an improved *Graphical User Interface* (GUI), along with better security. Table 5.2 shows the URLs and their supporting vendor for Beta Web Sites demonstrating UDDI Version 2. Data in these Beta UDDI Web sites may not be saved when they go operational.

TABLE 5.2   Beta Implementations of UDDI Version 2

Vendor	URL
Microsoft	*https://uddi.rte.microsoft.com/search/frames.aspx*
IBM	*https://www3.ibm.com/services/uddi/v2beta/protect/registry.html*
Hewlet Packard	*https://uddi.hp.com/uddi/index.jsp*
SAP	*https://websmp201.sap-ag.de/~form/uddi_discover*

NOTE

Notice that the URLs for UDDI Version 2 Beta use "HTTPS" instead of "HTTP." This means that your searches of these sites are encrypted, preventing others from knowing what you searched for. This is the same protection you receive at many E-commerce sites when you make a credit card transaction.

INTERNAL UDDI WEB SITES

A perhaps more useful implementation of UDDI involves installing an internal UDDI site at a large organization such as a telecom. A telecom could easily have hundreds of internal and external Web Services available to a large audience. By using UDDI, the corporations IT department takes advantage of a standard that many people are familiar with. In this situation, a new developer or consultant in the organization knows almost immediately where to find information about the services available in the organization.

If the repository is kept up to date, it could save an IT department money by reducing training and documentation costs.

Imagine that you're a consultant at a large corporation and you've been charged with bringing data on legacy systems, such as an IBM mainframe, to the public Web site.

Instead of hunting down an administrator, finding documentation, or having to create your own documentation, you simply go to the internal UDDI site. You search for the type of Web Service you're looking for, connect your Web page to this Web Service, and then deploy your new Web application.

Remember that a UDDI site only works if the people creating and using Web Services keep it up to date. All the advantages of UDDI are lost if users are not disciplined about disseminating the information.

A UDDI CASE STUDY

The examples so far in this chapter have been small in scope and they have not really represented the scenario UDDI was created to solve. UDDI's intent is to bring companies who want to do business over the Internet together quicker and in an automatic way.

Imagine that you want to set up a Web site to provide your users with meteorological information. Your first step is to identify the information you wish to provide and how you wish to present it. For example, you may wish to provide the temperature, forecast, and barometric pressure based on zip code. For the forecasts, you may want to provide moving images from satellites as a way to enhance your site.

Now that you've considered the information you want to provide, you need to determine how you want to display the information. In this case, you decide to provide your meteorological information in the form of Web Services, and then you create Web pages that call the services. By providing Web Services, you gain greater flexibility because Web pages (which may not even be on your Web site), wireless devices, and applications can consume the service. UDDI helps by providing a standard means of promoting your services.

Some of the information you provide comes from the home-built weather station in your back yard. The satellite imagery needs to come from another source; by doing a search on a UDDI site you find a Web Service for these images. The UDDI site provides information on where the URL that has the service resides, as well as any contact information needed in case the service requires a fee.

Once you complete your site with the Web Services, you can go back to the UDDI registry and create an entry. Currently, using the UDDI repository for promoting services you create is free, but as the repositories grow and cost more money to maintain this could change. Figure 5.8 illustrates how you can use a UDDI repository to search for the services needed and then use it to promote the resulting site.

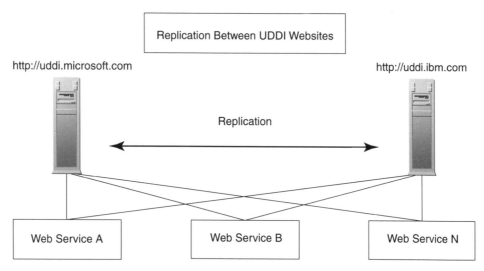

FIGURE 5.8   Illustrates how you can use a UDDI site to find the Web Services you need and then use the UDDI site to promote the resulting services.

Another aspect to consider is that the Web Service you search for may charge you money each time you use them, so you would want to pass these charges onto your customers. In the case of meteorological services, government agencies often have services available (although not always Web Services) to gather this information for free.

DON'T FORGET THE XML

Just as with the *SOAP* standard, UDDI describes multiple aspects—including how UDDI operates and how the XML underneath the information is presented. The information on UDDI sites is stored in a specific XML format. To see one of these

XML files, return to *http://uddi.microsoft.com* and do a search on business name for "Sun Microsystems." The links that come up are for Sun and one of its distributors. Click on the link for Sun and the results should look like Figure 5.9.

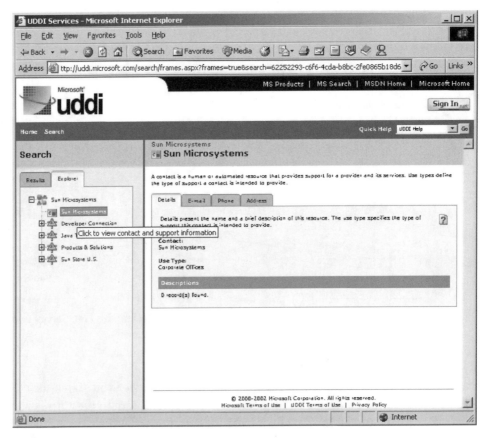

FIGURE 5.9   The results for searching for Sun Microsystems on *http://uddi.microsoft.com*.

Notice that the contact information for Sun is under "Business Details" and the contact information is just underneath that. If you scroll all the way down to the bottom of the page, you will find a subtitle labeled "Discovery URL." Underneath that you'll find the following URL: *http://www3.ibm.com/services/uddi/uddiget?businessKey=F293EE60-8285-11D5-A3DA-002035229C64*. This is the URL that represents Sun's UDDI XML data. Click on that link and you'll see some of the following XML code.

```xml
<?xmlversion="1.0" encoding="utf-8" ?>
<businessDetail generic="1.0" xmlns="urn:uddi-org:api"
                operator="www.ibm.com/services/uddi"
                 truncated="false">

<businessEntity authorizedName="1000000C5S"
   operator="www.ibm.com/services/uddi"
    businessKey="F293EE60-8285-11D5-A3DA-002035229C64">

<discoveryURLs>
 <discoveryURLuseType="businessEntity">http://www-3.ibm.com/
   services/uddi/uddiget?businessKey=F293EE60-8285-11D5-A3DA-
   002035229C64</discoveryURL>
 </discoveryURLs>
 <name>Sun Microsystems</name>
 <description xml:lang="en">Leading provider of
    industrial-strength hardware, software and
    services that power the Net</description>

<contacts>
       <contact useType="Corporate Offices">
    <personName>Sun Microsystems</personName>
<address>
   <addressLine>901 San Antonio
     Road</addressLine>
   <addressLine>Palo Alto</addressLine>
   <addressLine>CA 94303</addressLine>
   <addressLine>USA</addressLine>
 </address>
 </contact>
 </contacts>

 <businessServices>
<businessService serviceKey="02D4B1A0-8291-11D5-A3DA-002035229C64"
  businessKey="F293EE60-8285-11D5-A3DA-002035229C64">
 <name>Products & Solutions</name>
 <description xml:lang="en">Sun's Hardware & Software
   solutions</description>
<bindingTemplates>
```

```
<bindingTemplate bindingKey="62A3FA00-8291-11D5-A3DA-002035229C64"
  serviceKey="02D4B1A0-8291-11D5-A3DA-002035229C64">
<accessPoint URLType="http">http://www.sun.com/products-n-solutions/
  </accessPoint>
<tModelInstanceDetails>
<tModelInstanceInfo tModelKey="UUID:68DE9E80-AD09-469D-8A37-
  088422BFBC36" />
</tModelInstanceDetails>
</bindingTemplate>
</bindingTemplates>
</businessService>

    ...and many more busineesService entries
  </businessServices>
</businessDetail>
```

Note that the root element is businessDetail. It contains the operator attribute, which indicates where the entry is hosted. The businessEntity element contains information relevant to the corporation itself. The businessKey attribute is a unique identifier that is created by the UDDI Web site when the entry is made. The name element shows that this entry is for Sun Microsystems, the following element, description, provides information on the corporation, and then several elements give contact information.

The binding keys, such as 62A3FA00-8291-11D5-A3DA-002035229C64, *are unique identifiers that are created by the UDDI Web site when the entry is created.*

The businessServices element contains several child elements that are specific to a particular business service. Note that it doesn't necessarily describe a Web Service; in this case, the service entry brings you to a URL on Sun's Web site that describes their products and solutions in general.

Luckily, by using the GUIs on the UDDI sites, you can generate this XML just by entering some data into a Web form or by using the UDDI SDK provided by Microsoft and described in Appendix E.

CONCLUSION

This chapter introduces you to the concepts behind UDDI. Just like the *SOAP* standard, UDDI is a combination of communication and XML standard. UDDI Web sites such as *uddi.microsoft.com* allow you to search for Web Services and corporations with which you may want to do business. Internal UDDI sites can guide you to the Web Services you need to use within your corporation.

An important aspect of UDDI is that it only works as well as the people using it. If developers are not disciplined about entering information as Web Services come online, consumers can't discover the information. In fact, at the time of this writing there was not much information on any of the UDDI Web sites and often the information that was available was inconsistent. Hopefully as more Web Services come online, the amount of information in the repositories will increase and the quality and consistency will go up.

II Web Service Implementations

Section I covered the underlying theories and descriptions of how Web Services are required to behave. By this point you should have a good understanding of how all the different XML standards fit together.

Section II covers implementations of Web Services in C# and Java. You will see all the standards described in Section I put to use in these implementations. This should give you some ideas for the practical applications of Web Services.

6 .*NET* Web Services with C#

In This Chapter

- Prerequisites for Using .*NET* Web Services
- Creating .*NET* Web Services
- Simple C# Web Service and *Visual Studio.NET*
- Creating Web Services with the .*NET Framework SDK*
- Creating Web Service Consumers
- Discovery and .*NET*

B ecause Microsoft was involved early on in the development of the *SOAP* standard, its technology is light years ahead of much of the competition. The Web Service technology that comes with Microsoft's *.NET* environment is easy to use, is intuitive, provides easy methods of documentation, and possesses several methods of discovery.

Perhaps the best part of the Microsoft implementation is that you don't have to purchase *Visual Studio.NET* in order to implement or distribute this implementation. Microsoft has made the *.NET Framework SDK™* available free from their Web site. This gives you command line compilers for C# and *Visual Basic*, along with the ability to create *ASP.NET* pages in either language. Many of the examples in this chapter show you how to use the *.NET Framework SDK* to compile code and the *ASP.NET* pages. Therefore, it is not necessary to own *Visual Studio.NET* to execute and compile the examples.

There is also the *.NET Framework Redistributable™*, which sets up a server or client to support the distribution of *.NET* tools. If you have experience with Java, the *.NET Redistributable* is very similar to the Java *Runtime Environment™*. The *Redistributable* sets up a machine to run *.NET* applications, *ASP.NET* Web pages, and Web Services, but does not support compilation.

This chapter covers many aspects of Web Services in *.NET*. These aspects include making a simple Web Service and then looking at the various consumers, including Web pages and applications. The prerequisites needed on your machine are also covered in this chapter.

PREREQUISITES FOR USING *.NET* WEB SERVICES

To use either *Visual Studio.NET* or the *.NET Framework SDK*, you will need to have *Internet Information Server* (IIS) installed on your system. This means that you'll need to be using Microsoft *Windows NT 4.0*, *Windows 2000*, or *Windows XP Professional*. Unfortunately, IIS and *.NET* do not run on the consumer *Windows* installations such as *Windows ME* or *Windows 98*.

To install the Web server (and note that this procedure works on all three supported operating systems), go to "Start" and then "Control Panel." Select "Add/ Remove Programs."

On the lefthand side there is an icon that says "Add/Remove Windows Components."

Click on that icon. Figure 6.1 shows the "Add/Remove Windows Components" dialogue on *Windows 2000*.

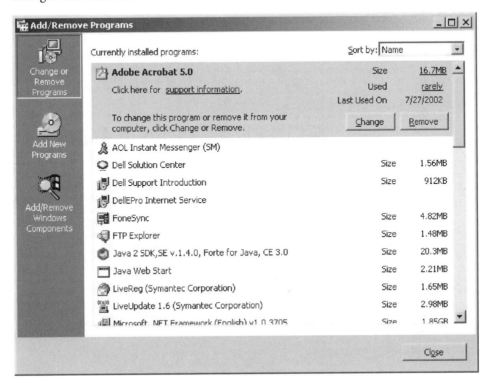

FIGURE 6.1 The "Add/Remove Programs" dialogue for *Windows 2000*. Notice the "Add/Remove Windows Components" on the lefthand side.

Figure 6.2 shows the "Add/Remove Windows Components" dialogue for *Windows 2000*. Note that the box for "Internet Information Services" is already checked for this system. This means that the server is already installed in this case.

If this box is unchecked, check it and insert the CD-ROM for the operating system installed on your computer and click "Next." The dialogue that comes up guides you through the installation of IIS. It will ask you to choose a home directory for a Web site on your system. You should choose a directory that's convenient for you. On the author's system, the root directory is *c:\inetpub*.

Once IIS is installed, you can install either *Visual Studio.NET* or the *.NET Framework SDK*. The instructions for installing these programs are straightforward and come with installation programs, but you may want to move one *MS DOS Batch* (bat) file to someplace convenient. *Corvars.bat* sets up the DOS environment on your PC

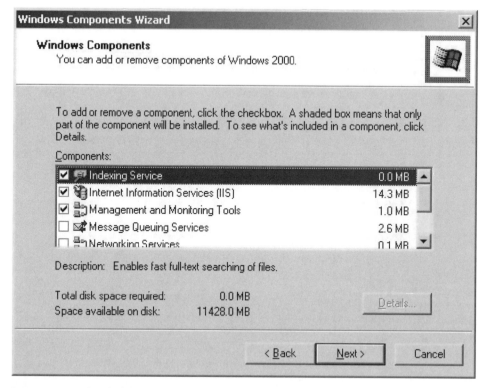

FIGURE 6.2   The dialogue to select which *Windows* components to install.

so that you can use the *.NET framework SDK* to compile the various code examples in this chapter on your PC. You can set up the installation programs for either *.NET* product to set these for you automatically, but then they are set all the time and can create a very complex environment on your PC. By allowing the bat file to set this environment for you, you have a convenient way of using the *.NET* command line tools and the environment vars are only there if you need them. Once IIS and the appropriate *.NET* environment are installed, you're ready to begin writing Web Services.

CREATING *.NET* WEB SERVICES

The Microsoft implementation of creating Web Services is very advanced, which makes it easy for a developer to create and deploy these Web Services. By providing an interface for *Internet Explorer*, the services are easy to test and document. You'll also find in this chapter that applications are not the only consumers of Web Services. Web pages, whose content can be customized for a variety of devices including cell phones or PDAs that have access to the Web, can also be consumers of Web Services.

This chapter starts off with creating a simple stock quote Web Service, and then you will create several different consumers of that service.

SIMPLE C# WEB SERVICE AND *VISUAL STUDIO.NET*

Open *Visual Studio.NET*. From the "File" menu, select "New Project." On the left-hand side, select "Visual C# Projects" and on the righthand side select "ASP.NET Web Service."

Next, select the location under your Web root in which you want the Web Service to appear. Give the server an appropriate name such as "SimpleStockQuote." Figure 6.3 displays the project selection dialogue box.

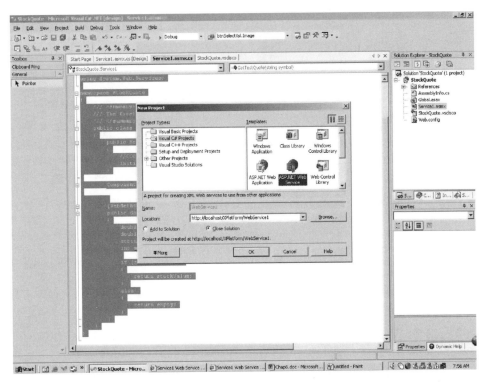

FIGURE 6.3   A demonstration of how to create a C# Web Service project in *Visual Studio.NET*.

Once you name the service, *Visual Studio.NET* brings you into design mode, which looks like Figure 6.4. Look for the link in the center of the screen to be brought into the code window.

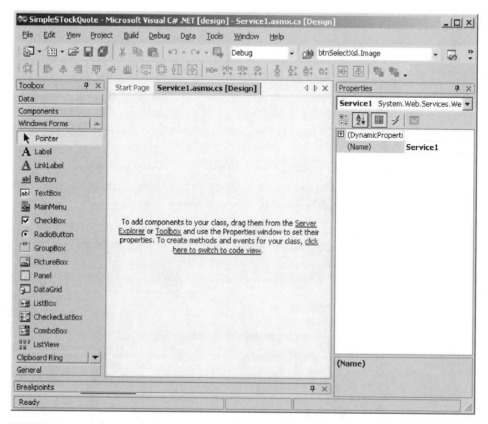

FIGURE 6.4 Design mode for creating Web Services in *Visual Studio.*

Once you are in the code window, there is a lot of code already generated for you. The following code sample shows this generated code.

The first things to notice in the sample are all the "using" statements. These are the namespaces (known as "includes" in C++ or "imports" in Java) that Microsoft uses to make the services possible. These namespaces are used by the code generated for you; you don't really need to worry about them.

Next there's the namespace definition for this service. Right now this code sample uses the default that Microsoft assigned, which is WebService1. You can ignore much of the rest of the code because it initializes and destroys the objects that work behind the scenes to make the Web Service possible.

Notice the `hello world` example Microsoft included in the code.

```csharp
using System;
using System.Collections;
using System.ComponentModel;
using System.Data;
using System.Diagnostics;
using System.Web;
using System.Web.Services;

namespace WebService1
{
  /// <summary>
  /// Summary description for Service1.
  /// </summary>
  public class Service1 : System.Web.Services.WebService
  {
      public Service1()
      {
            //CODEGEN: This call is required by the ASP.NET Web
                      Services Designer
            InitializeComponent();
      }

      #region Component Designer generated code

      //Required by the Web Services Designer
      private IContainer components = null;

      /// <summary>
      /// Required method for Designer support - do not modify
      /// the contents of this method with the code editor.
      /// </summary>
      private void InitializeComponent()
      {
      }

      /// <summary>
      /// Clean up any resources being used.
      /// </summary>
      protected override void Dispose( bool disposing )
```

```
        {
            if(disposing && components != null)
            {
                components.Dispose();
            }
            base.Dispose(disposing);
        }

        #endregion

        // WEB SERVICE EXAMPLE
        // The HelloWorld() example service returns the string
        //   Hello World
        // To build, uncomment the following lines then save and
        //build the project
        // To test this Web Service, press F5

        //[WebMethod]
        //public string HelloWorld()
        //{
        //      return "Hello World";
        //}
    }
}
```

The stock quote example in this chapter really won't look up stock quotes, but will just provide you with a value for one particular symbol and then a negative number for all other values. The examples in this chapter are simple and meant only to introduce you to the basic concepts.

To modify the service Microsoft creates for you, first change the namespace to something that is appropriate for your environment. In this example, the namespace chosen is StockQuote. Then skip down to where the code states [WebMethod]. You can comment out the code at that point and add the following code.

```
[WebMethod]
public double GetTestQuote(string symbol)
{
  double stockValue = 55.95;
  double empty = -1;

  if (symbol == "C")
  {
```

```
      return stockValue;
    }
    else
    {
     return empty;
    }
  }
```

Notice that in the preceding code example the only real difference between this code and the code example in Chapter 1 is the addition of [WebMethod] to the definition of the method. Other than that, the code is a standard C# method.

The code simply returns a numeric value or price for the symbol c and -1 for every other symbol passed to it.

After adding the last code snippet to your Web Services code, the entire Web Service code should now look like the following.

```
using System;
using System.Collections;
using System.ComponentModel;
using System.Data;
using System.Diagnostics;
using System.Web;
using System.Web.Services;

//modified namespace --Xplatform book
namespace StockQuote
{
  /// <summary>
  /// The first C# Web Service
  /// </summary>
  public class Service1 : System.Web.Services.WebService
  {
        public Service1()
        {
              //CODEGEN: This call is required by the ASP.NET Web
              //Services Designer
              InitializeComponent();
        }

        #region Component Designer generated code

        //Required by the Web Services Designer
```

```csharp
private IContainer components = null;

/// <summary>
/// Required method for Designer support - do not modify
/// the contents of this method with the code editor.
/// </summary>
private void InitializeComponent()
{
}

/// <summary>
/// Clean up any resources being used.
/// </summary>
protected override void Dispose( bool disposing )
{
    if(disposing && components != null)
    {
        components.Dispose();
    }
    base.Dispose(disposing);
}

#endregion


[WebMethod]
public double GetTestQuote(string symbol)
{
    double stockValue = 55.95;
    double empty = -1;

    if (symbol == "C")
    {
        return stockValue;
    }
    else
    {
        return empty;
    }
}
}
}
```

Under "Debug" in *Studio*, click on "Start without Debugging," which is symbolized with "!". This executes the Web Service and opens a browser window that looks like Figure 6.5.

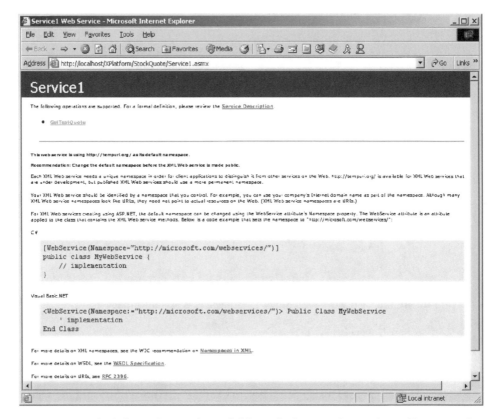

FIGURE 6.5 The information made available to the browser from Microsoft's *.NET* Web Services implementation.

The first link available is the link to the service description, which is the WSDL file. Any Microsoft Web Services that have the "asmx" extension simply need WSDL passed as the query string in order to see the description. In this case, this URL gives us the WSDL file: *http://localhost/XPlatform/StockQuote/Service1.asmx?WSDL*. Figure 6.6 shows the WSDL information in *Internet Explorer*.

Remember WSDL describes the Web Service so a consumer, such as a Web page or application, knows what methods and variables are available. This is used by .NET to create a proxy. Chapter 4 covers WSDL in detail.

NOTE

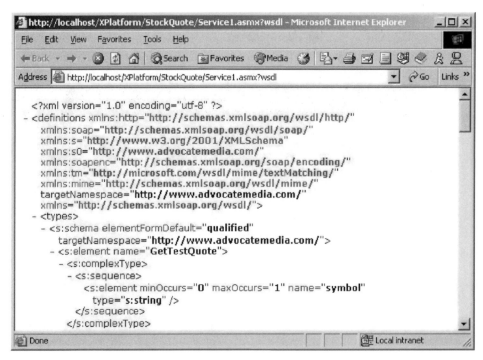

FIGURE 6.6   How the Web Service displays its WSDL information.

After the link for WSDL, there is a link for the one method available in this service and that's GetTestQuote. This link reveals methods of testing the service. Figure 6.7 displays how this output looks in the browser.

The information on the method page shows you the *SOAP* requests and response, along with a dialogue to test the implementation. For the simple stock quote example, the request looks like the following:

```
POST /XPlatform/StockQuote/Service1.asmx HTTP/1.1
Host: localhost
Content-Type: text/xml; charset=utf-8
Content-Length: length
SOAPAction: "http://tempuri.org/GetTestQuote"

<?xml version="1.0" encoding="utf-8"?>
<soap:Envelope
 xmlns:xsi="http://www.w3.org/2001/XMLSchema-instance"
```

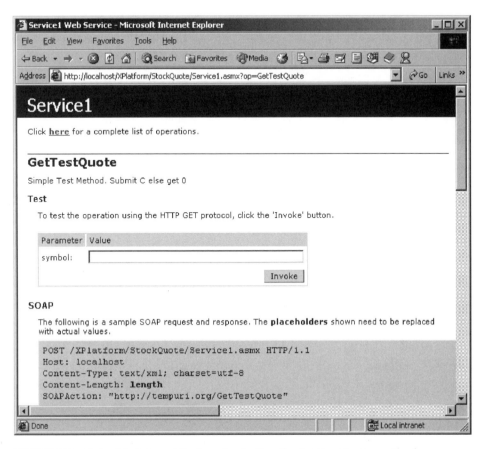

FIGURE 6.7 Output in *Internet Explorer* for testing the `GetTestQuote` method.

```
xmlns:xsd="http://www.w3.org/2001/XMLSchema"
xmlns:soap="http://schemas.xmlsoap.org/soap/envelope/">
<soap:Body>
<GetTestQuote xmlns="http://tempuri.org/">
  <symbol>string</symbol>
</GetTestQuote>
</soap:Body>
</soap:Envelope>
```

Notice that in this example, the Web Service leaves out things like the actual length and the value being sent. Both of these values vary depending on the request you make.

Note that if you're calling this Web Service from another language, such as Java, you could send the request to the address of the Web Service and get the following response back.

```
:
HTTP/1.1 200 OK
Content-Type: text/xml; charset=utf-8
Content-Length: length

<?xml version="1.0" encoding="utf-8"?>
<soap:Envelope
 xmlns:xsi="http://www.w3.org/2001/XMLSchema-instance"
 xmlns:xsd="http://www.w3.org/2001/XMLSchema"
xmlns:soap="http://schemas.xmlsoap.org/soap/envelope/">
  <soap:Body>
    <GetTestQuoteResponse xmlns="http://tempuri.org/">
      <GetTestQuoteResult>double</GetTestQuoteResult>
    </GetTestQuoteResponse>
  </soap:Body>
</soap:Envelope>
```

Remember that the SOAP standard describes how to encapsulate data in XML during transmission to a node along with the protocols, such as HTTP, that move the data to the appropriate node. For more information, refer back to Chapter 3.

If you enter the symbol "C" into the text box on this page, you'll get the following response in *Internet Explorer*.

```
<?xml version="1.0" encoding="utf-8" ?>
<double xmlns="http://tempuri.org/">55.95</double>
```

Note that *tempura.org* is the default namespace that *Visual Studio.NET* defines.

CREATING WEB SERVICES WITH THE .NET FRAMEWORK SDK

If *Visual Studio.NET* is unavailable to you, you can easily download the *.NET Framework SDK* from Microsoft. This provides you with command line compilers for C# and *Visual Basic*, along with all the libraries to create *ASP.NET* pages and Web Services.

You can take the code from the previous example, put it in a directory under your Web root, and name the file with an ".asmx" extension on the end. Then, by simply pointing a browser to the Web Service, the code compiles.

There is one minor modification to be made to the code that involves adding the following line to the beginning of the page.

```
<%@ WebService Language="C#" Class="StockQuote.XPlatformServices" %>
```

This tag indicates to *ASP.NET* that this is a Web Service written in C#. Just to ensure that you have the entire picture of how Web Service codes need to look, the following is the entire code listing with the *ASP.NET* tag included.

```
<%@ WebService Language="C#" Class="StockQuote.XPlatformServices"
%>
using System;
using System.Collections;
using System.ComponentModel;
using System.Data;
using System.Diagnostics;
using System.Web;
using System.Web.Services;

//modified namespace --Xplatform book
namespace StockQuote
{
  /// <summary>
  /// The first C# Web Service
  /// </summary>
  public class Service1 : System.Web.Services.WebService
  {
      public Service1()
      {
            //CODEGEN: This call is required by the ASP.NET Web
            //Services Designer
            InitializeComponent();
      }

      #region Component Designer generated code

      //Required by the Web Services Designer
      private IContainer components = null;

      /// <summary>
```

```
/// Required method for Designer support - do not modify
/// the contents of this method with the code editor.
/// </summary>
private void InitializeComponent()
{
}

/// <summary>
/// Clean up any resources being used.
/// </summary>
protected override void Dispose( bool disposing )
{
        if(disposing && components != null)
        {
                components.Dispose();
        }
        base.Dispose(disposing);
}

#endregion


[WebMethod]
public double GetTestQuote(string symbol)
{
        double stockValue = 55.95;
        double empty = -1;

        if (symbol == "C")
        {
                return stockValue;
        }
        else
        {
                return empty;
        }
}
}
}
```

When *Visual Studio.NET* generates a C# Web Service, it separates the C# code from the *ASP.NET* page with an additional attribute in the WebService tag. Consider the following example.

```
<%@ WebService
 Language="c#"
 Codebehind="Service1.asmx.cs"
 Class="StockQuote.XPlatformServices" %>
```

The Codebehind attribute imports the C# code into the *ASP.NET* page. This is all the code that's needed for *ASP.NET*.

Documenting the Service

Microsoft added features to allow you to provide documentation to users who browse with *Internet Explorer*. The WebService and WebMethod elements in the C# code allow you to customize the output of the browser and set a namespace that is unique for your project.

The namespace is just a unique identifier for a particular project. By setting it to something useful to you, you eliminate the message from Microsoft that would have told you to change the namespace. You also make the service unique to your project.

The WebService attribute provides a method of creating a description and a unique namespace by simply adding the following code about the class definition.

```
[WebService(Description="Web Services for XPlatform Book",
            Namespace="http://www.advocatemedia.com/")]
```

The WebMethod attribute, which defines a method as being part of a Web Service, also allows you to provide extra information to the browser simply by adding the following code.

```
[WebMethod(Description="Simple Test Method. Submit C else get -1")]
```

By adding this extra information you provide the user a better view to what the service does. Figure 6.8 shows how the browser displays this information.

Changing the name of the class to XPlatformServices makes the Web Service slightly more descriptive and this ends up in the Web page as well.

The following complete code example shows the WebService and WebMethod tags in the proper positions.

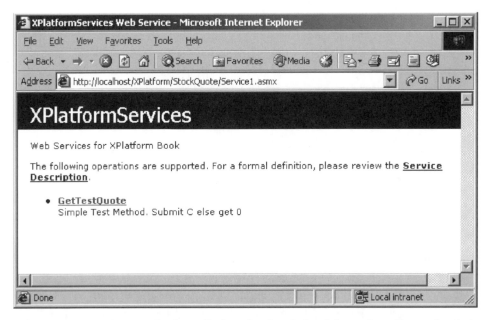

FIGURE 6.8 How *Internet Explorer* displays the description information. Also notice that the warning from Microsoft that you need to change the namespace is no longer present.

```csharp
using System;
using System.Collections;
using System.ComponentModel;
using System.Data;
using System.Diagnostics;
using System.Web;
using System.Web.Services;

namespace StockQuote
{
  /// <summary>
  /// Web Services for XPlatform Web Services Book
  /// </summary>
  [WebService(Description="Web Services for XPlatform Book",
              Namespace="http://www.advocatemedia.com/")]
  public class XPlatformServices: System.Web.Services.WebService
```

```csharp
{
    public XPlatformServices()
    {
        //CODEGEN: This call is required by the ASP.NET Web
        //Services Designer
        InitializeComponent();
    }

    #region Component Designer generated code

    //Required by the Web Services Designer
    private IContainer components = null;

    /// <summary>
    /// Required method for Designer support - do not modify
    /// the contents of this method with the code editor.
    /// </summary>
    private void InitializeComponent()
    {
    }

    /// <summary>
    /// Clean up any resources being used.
    /// </summary>
    protected override void Dispose( bool disposing )
    {
        if(disposing && components != null)
        {
            components.Dispose();
        }
        base.Dispose(disposing);
    }

    #endregion

    [WebMethod(Description="Simple Test Method. Submit C else
        get -1")]
    public double GetTestQuote(string symbol)
    {
        double stockValue = 55.95;
```

```
                      double empty = -1;


                      if (symbol == "C")
                      {
                            return stockValue;
                      }
                      else
                      {
                            return empty;
                      }
               }
         }
   }
```

CREATING WEB SERVICE CONSUMERS

Web Service consumers actually utilize the functionality found in the methods of a particular service. A consumer may be a *Windows* application, a Web page, or any device that can utilize the *.NET* environment. In this section, the consumers include a simple *Windows* application and a Web page. This section also explores utilizing the *.NET Framework SDK* without the aid of *Visual Studio.NET* to create consumers.

Creating a Windows Consumer with *Visual Studio.NET*

To utilize a Web Service from a *Windows* program, open *Visual Studio.NET* and select "Visual C# Projects" on the lefthand side of the "New Project" dialogue box. On the righthand side select "Windows Application." Figure 6.9 shows the proper selections in the "New Project" dialogue.

Go into design mode of *Visual Studio* and make the *Windows* application look like what appears in Figure 6.10. This involves dragging and dropping two text boxes, two labels, and a button onto the *Windows* form. Name the top text box "simpleSymbol" and the bottom text box "stockResult." The button needs to be named "testButton."

Figure 6.10 shows the design of the *Windows* program so far.

Next, from the project menu, select "Add Web Reference." This brings up the "Add Web Reference" dialogue shown in Figure 6.11.

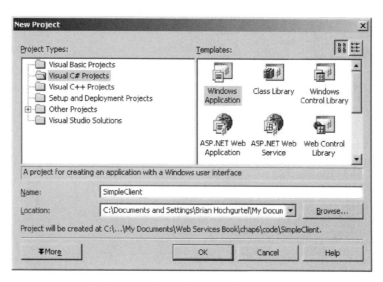

FIGURE 6.9   The "New Project" dialogue box for creating a *Windows* application.

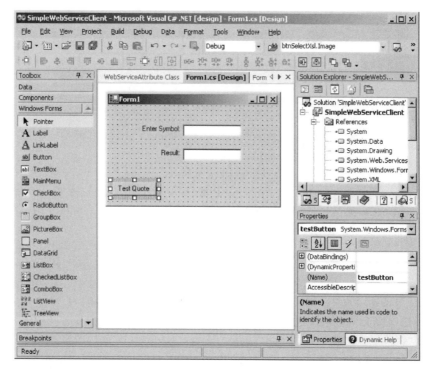

FIGURE 6.10   The design of the *Windows* program.

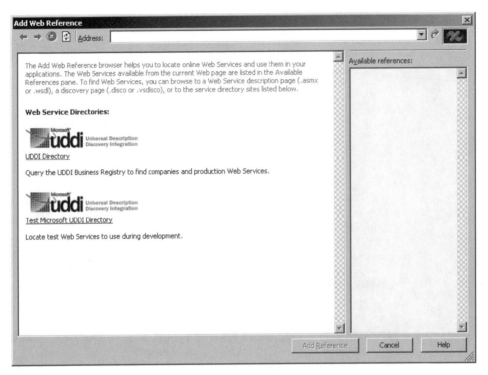

FIGURE 6.11 The "Add Web Reference" dialogue of *Visual Studio.NET*.

To create the Web reference, you must enter the path of the WSDL file through a URL and not a local file.

The URL used in this example is: *http://localhost/XPlatform/StockQuote/Service1.asmx?wsdl*. Your path may vary depending on where you installed your Web root and Web Service.

If you don't know the URL of your Web Service, open the service's project in Visual Studio.NET *and execute it without debugging. This will open a browser window to the service. Then click the service description link. The URL displayed in the address bar of your browser now reveals the WSDL needed to make the Web reference.*

Figure 6.12 shows how the Web reference to local host appears in the "Solution Explorer" window of *Visual Studio.NET*.

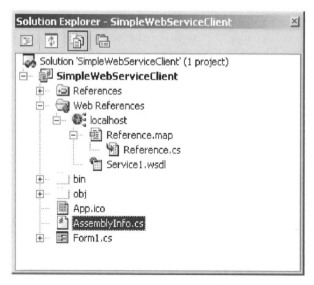

FIGURE 6.12 The Web reference to local host in the "Solution Explorer" window.

When the project is built, the Web reference creates a dll for the project to access the functionality in the Web Service. For an in-depth explanation of what happens in this step, see "Creating Consumers with the *.NET Framework SDK*" later in this chapter. This section examines doing this step manually and should help you understand what *Visual Studio.NET* does for you when it creates a reference.

The next step is to double click on "testButton" in the window designer (see Figure 6.10).

This brings up the click event for the button, and here is where you need to put the code for connecting the Web Service.

The following code sample calls the GetTestQuote method described when you created the Web Service from the first example in this chapter. This code first defines two strings for sending and receiving values. The variable stockSymbol gets its value from the enterSymbol text field. Then the following line of code creates the example object.

```
localhost.XPlatformServices example = new
localhost.XPlatformServices();
```

Localhost is the name of the site the Web reference was created for and XplatformService is the name of the service for which the Web reference was created.

This is how you create objects in C#. By creating the Web reference, you can treat the Web Service as if it were any other object.

Now that the example object is created, the code calls the GetTestQuote method, which always returns the value as a string. This happens because of the ToString() attached on the end of the original method call. Then the stockResult text field receives the value of the stockValue variable. Note that stockResult.Text sends the value of stockValue to the field on the form.

```
private void testButton_Click(object sender, System.EventArgs e)
    {
    //stock value returned by the service
    string stockValue;

    //the symbol entered into the enterSymbol text box
    string stockSymbol = enterSymbol.Text;

    //create object example to communicate with service
    localhost.XPlatformServices example = new
    localhost.XPlatformServices();

    //call the GetTestQuote method convert to a string for display
    stockValue = example.GetTestQuote(stockSymbol).ToString();

    //Put the result in the appropriate text box.
    stockResult.Text = stockValue;
    }
```

To execute the example, select "Build" from the toolbar in *Visual Studio.NET* and "Build Solution," because you need to compile the Web reference in addition to the *Visual C#* code. If it builds without errors, select "Debug" and then "Start." This executes the *Windows* application you just created. Figure 6.13 shows the *Visual C#* application executing.

Remember that the Web Service only returns a value for the symbol "C" whereas it returns -1 for everything else.

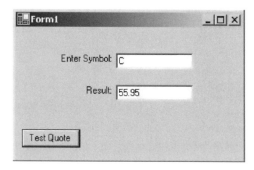

FIGURE 6.13   The *Visual C#* client executing
and receiving a value back from the Web Service.

ASP.NET WEB PAGE WITH *VISUAL STUDIO.NET*

An *ASP.NET* page runs as a Web page on your server that is viewable with *Internet Explorer*. Because code runs every time an *ASP.NET* page loads, it is possible to call objects like any other C# program.

Open *Visual Studio.NET* and select *Visual C#* and select *ASP.NET Web Application* in the dialogue. Be sure to provide the application with a name that's meaningful to you. Figure 6.14 shows the dialogue and the appropriate selections.

Visual Studio.NET creates a Web page that looks like the following.

```
<%@ Page language="c#" Codebehind="WebForm1.aspx.cs"
AutoEventWireup="false" Inherits="SimpleASPClient2.WebForm1" %>

<!DOCTYPE HTML PUBLIC "-//W3C//DTD HTML 4.0 Transitional//EN" >
<html>
  <head>
    <title>WebForm1</title>
    <meta name="GENERATOR" Content="Microsoft Visual Studio 7.0">
    <meta name="CODE_LANGUAGE" Content="C#">
    <meta name="vs_defaultClientScript" content="JavaScript">
    <meta name="vs_targetSchema"
          content="http://schemas.microsoft.com/intellisense/ie5">
  </head>
  <body MS_POSITIONING="GridLayout">
    <form id="Form1" method="post" runat="server">
    </form>
```

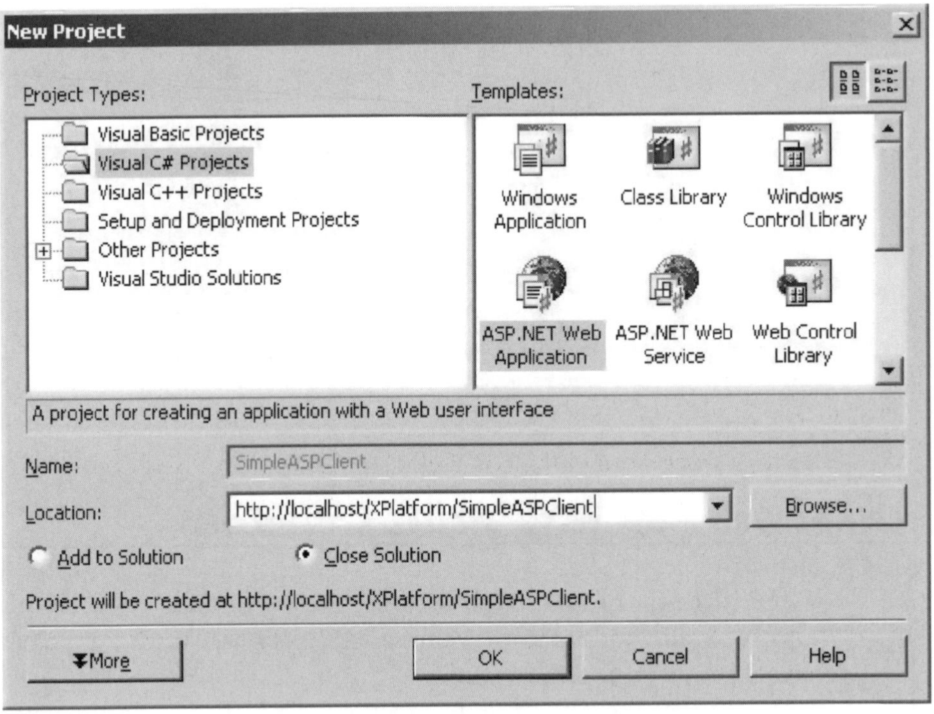

FIGURE 6.14 The appropriate selections for creating an *ASP.NET Web Application* based on C#.

```
        </body>
    </html>
```

A lot of the information here is extraneous and can be trimmed down to save space. The `meta` tags are just information about the Web page and don't really affect the code. The DOCTYPE definition simply defines the HTML used in the page, and the MS_POSITIONING attribute of the body tag is a directive for *Internet Explorer* that we won't use. Thus, the code can be trimmed down to the following.

```
<%@ Page language="c#"
        Codebehind="WebForm1.aspx.cs"
        AutoEventWireup="false"
        Inherits="SimpleASPClient2.WebForm1" %>

<html>
```

```
<head>
    <title>Simple ASP.NET Web Service Client</title>
</head>
<body>
    <form id="Form1" method="post" runat="server">
    </form>
</body>
</html>
```

The next step is to add a Web reference to the Web Service on localhost that you used in the previous sections' examples. Go to "Project" in *Visual Studio.NET* and select "Add Web Reference." Put in the URL for the WSDL file for the "Simple-StockQuote" Web Service and the reference will be built with the project. See Figure 6.11 for a description of the Web Reference dialogue in *Visual Studio.NET*.

Once *Visual Studio.NET* creates the Web reference, the code now contains an import statement that brings in the namespace for the Web Service. The code looks like the following.

```
<%@ Import Namespace="SimpleASPClient.localhost" %>
```

This imports in the localhost namespace so that when you create objects from the localhost Web Service, the definitions aren't as long.

The second tag found in the page, Page, defines the language used, the location of the auto-generated code, and what code is inherited. *Visual Studio.NET* creates much of the information here from values you entered when creating the *ASP.NET* project. The AutoEventWireup attribute indicates whether the events on the Web page fire off when the page is first loaded. Inherits the compiler the location of the code that *Visual Studio.NET* created for this *ASP.NET* page.

After the HTML tag, the script tag appears. The attribute, runat="server", indicates to the Web server that this is a piece of code that needs to be executed on the server side. Within the script tags is that actual chunk of code that calls the Web Service.

The first step is creating a method named Submit_Click. This method will be called when someone clicks on a submit button, which is defined later. Note the arguments to the Submit_Click method (Object Src, EventArgs E) actually load the Web reference created earlier.

Next the object example gets created to call the methods in the Web Service.

Note that you don't have to use the localhost namespace in this definition because of the import directive used earlier.

The string value, stockSymbol, receives its value from the *ASP.NET* control SymbolToGet. This is simply a text box dragged and dropped onto the design window in *Visual Studio.NET*. stockValue is defined as a string and set to a default of empty and then it receives the value back from the GetTestQuote method. Notice that the ToString() method call ensures that the value returned is a string and, thus, can be sent to the Text attribute of the Message TextBox.

The following is the complete code.

```
<%@ Import Namespace="SimpleASPClient.localhost" %>
<%@ Page language="c#"
    Codebehind="WebForm1.aspx.cs"
    AutoEventWireup="false"
    Inherits="SimpleASPClient.WebForm1" %>
<HTML>
  <script runat="server">
  protected void Submit_Click(Object Src, EventArgs E)
  {
    XPlatformServices example = new XPlatformServices();
    string stockSymbol = SymbolToGet.Text;
    string stockValue = "";
    stockValue = example.GetTestQuote(stockSymbol).ToString();
    Message.Text = stockValue;
   }
  </script>
  <body>
  <form runat="server">
    <b>Stock Symbol:</b>
        <asp:TextBox ID="SymbolToGet" Runat="server" /><br>
    <b>Returned:</b>
<asp:Label ID="Message" Runat="server" /><br>
        <input type="submit"
               value="get quote"
               onserverclick="Submit_Click"
               runat="server">
    </form>
    </body>
  </HTML>
```

Note that the asp:Textbox *and the* asp:Label *are available from the HTML tab of the "Tools" window in* Visual Studio.NET.

Figure 6.15 shows the output of the *ASP.NET* code in *Internet Explorer*. Note that by entering the symbol "C," we got back the value "55.95," just as in the previous Web Service consumer examples.

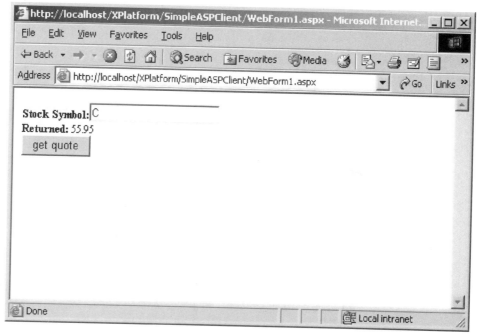

FIGURE 6.15   The output of the previous *ASP.NET* code in *Internet Explorer*.

There's really nothing special about the form we created to connect to our example Web Service. Thus, the example doesn't really take advantage of the customization that can occur with an *ASP.NET* page being able to utilize HTML. By just adding a few HTML directives, you can customize a user's experience with the *ASP.NET* page.

The following code example illustrates how you can customize the appearance in the browser by adding a few HTML tags. Note that all that has really been added is a title in the header, a table with a border and background color within the form code, and an HTML H3 definition (See Figure 6.16).

```
<%@ Page language="c#"
              Codebehind="WebForm1.aspx.cs"
```

```
                        AutoEventWireup="false"
                        Inherits="SimpleASPClient.WebForm1" %>
  <%@ Import Namespace="SimpleASPClient.localhost" %>
  <HTML>
  <head>
      <title>Simple ASP.NET Web Services consumer</title>
  </head>

  <script runat="server">
   protected void Submit_Click(Object Src, EventArgs E)
   {
       XPlatformServices example = new XPlatformServices();
       string stockSymbol = SymbolToGet.Text;

       string stockValue = "";
       stockValue = example.GetTestQuote(stockSymbol).ToString();
       Message.Text = stockValue;
   }
  </script>
  <body>
      <form runat="server">
      @h3:Simple ASP.NET Consumer</h3>

      <table align="center" bgcolor="#FFFCC" border=1>
        <tr>
            <td><b>Stock Symbol:</b></td>
            <td><asp:TextBox ID="SymbolToGet" Runat="server" /></td>
        </tr>
        <tr>
            <td><b>Returned:</b></td>
            <td><asp:Label ID="Message" Runat="server" /></td>
        </tr>
      </table>
    <input type="submit"
           value="get quote"
           onserverclick="Submit_Click" \
           runat="server"
           ID="Submit1"
           NAME="Submit1">
      </form>
    </body>
  </HTML>
```

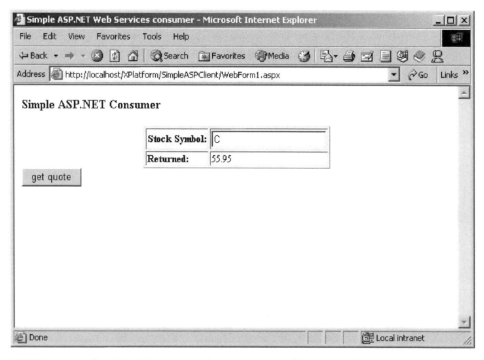

FIGURE 6.16   The *ASP.NET* Web Services consumer with some minor cosmetic changes.

Using an *ASP.NET* as a Web Services consumer provides for some easy customization for the user experience.

Using the *.NET* Framework SDK to Create Consumers

Just as with creating Web Services, you can use the *.NET Framework SDK* as a way to create Web Service consumers. *Visual Studio.NET* has a wide variety of features that allow you to easily create *Windows* and other applications, but if this version of *Visual Studio* is unavailable to you, the *.NET Framework SDK* allows you to create Web Services and Web Service consumers free of charge.

Creating the Proxy

With the *SDK*, you have to create the Web references, known as proxies in this case, manually. This is done by utilizing the WSDL tool and the command line C# compiler.

The first step is to use the WSDL tool to create the proxy code. Open a DOS prompt and ensure that you either execute the *corvars.bat* program that came with the *SDK* or that the DOS environment already has the appropriate environment set up for you. You can test this by executing *wsdl.exe*. If the DOS environment doesn't know about WSDL, that means you need to find and execute *corvars.bat*.

Note that the example command assumes that your StockQuote Web Service resides at the following URL: *http://localhost/XPlatform/StockQuote/Service1.asmx?wsdl*

Once your DOS environment is ready, go ahead and execute the following. (Note that you want all this information to be on one line.)

```
wsdl.exe /l:CS /n:SDKStocks /out:stockquotes.cs
http://localhost/XPlatform/StockQuote/Service1.asmx?wsdl
```

Table 6.1 explains the command line options for the WSDL tool.

TABLE 6.1 Options for the WSDL Tool from Microsoft

Command Line Option	Purpose
/l:CS	The language of the resulting proxy is C#.
/n:StockQuotes	Use the namespace StockQuotes in the proxy.
/out:stockquotes.cs	The resulting source file for the proxy will be stockquotes.cs.

Executing the command creates the stockquotes.cs file and the contents of it look like the following.

```
//------------------------------------------------------------
// <autogenerated>
//     This code was generated by a tool.
//     Runtime Version: 1.0.3705.0
//
//     Changes to this file may cause incorrect behavior and will be
//       lost if
//       the code is regenerated.
// </autogenerated>
```

```
//---------------------------------------------------------------

//
// This source code was auto-generated by wsdl, Version=1.0.3705.0.
//
namespace SDKStocks {
    using System.Diagnostics;
    using System.Xml.Serialization;
    using System;
    using System.Web.Services.Protocols;
    using System.ComponentModel;
    using System.Web.Services;


    /// <remarks/>
    [System.Diagnostics.DebuggerStepThroughAttribute()]
    [System.ComponentModel.DesignerCategoryAttribute("code")]
    [System.Web.Services.WebServiceBindingAttribute
        (Name="XPlatformServicesSoap",
         Namespace="http://www.advocatemedia.com/")]
         public class XPlatformServices :
    System.Web.Services.Protocols.SoapHttpClientProtocol {

    /// <remarks/>
    public XPlatformServices() {
        this.Url =
"http://localhost/XPlatform/StockQuote/Service1.asmx";
    }

    /// <remarks/>
    [System.Web.Services.Protocols.SoapDocumentMethodAttribute
      ("http://www.advocatemedia.com/GetTestQuote",
       RequestNamespace="http://www.advocatemedia.com/",
       ResponseNamespace="http://www.advocatemedia.com/",
       Use=System.Web.Services.Description.SoapBindingUse.Literal,
ParameterStyle=System.Web.Services.Protocols.SoapParameterStyle.Wrapped
      )]
    public System.Double GetTestQuote(string symbol) {
        object[] results = this.Invoke("GetTestQuote", new object[]
    {
                   symbol});
        return ((System.Double)(results[0]));
```

```
            }

            /// <remarks/>
            public System.IAsyncResult BeginGetTestQuote(string symbol,
            System.AsyncCallback callback, object asyncState) {
                return this.BeginInvoke("GetTestQuote", new object[] {
                            symbol}, callback, asyncState);
            }

            /// <remarks/>
            public System.Double EndGetTestQuote(System.IAsyncResult
asyncResult) {
                object[] results = this.EndInvoke(asyncResult);
                return ((System.Double)(results[0]));
            }
        }
    }
```

You shouldn't be intimidated by the code created for the proxy. An automated process creates this code and it isn't meant for you to modify or use directly.

You should never modify this code. Let the WSDL tool do that for you because the code gets overwritten every time the tool executes. Thus, you would lose your changes.

Now that you created the proxy code, it needs to be compiled so that the application or Web page can utilize it. You compile it from the command line, as shown in the following example. Note that the line breaks are present to make it easier for you to read.

```
    csc /t:library /out:XPlatform.dll /r:system.dll
/r:system.weservices.dll /r:system.xml.dll stockquotes.cs
```

Remember that in order to use the C# command line compiler, csc, you either need to have the appropriate environment already set up or you need to execute *corvars.bat*. The command line options are very important in this step and Table 6.2 summarizes them.

Because you don't have *Visual Studio.NET's* Graphical User Interface (GUI) to handle many of these details, you must use the command line options to handle

TABLE 6.2　A Summary of the Options Used for the C# Compiler to Build the Proxy

Command Line Option	Purpose
/t:library	Tells the compiler you want to create a dll.
/out:Xplatform.dll	Sets the name of the dll.
/r:system.dll	Has the dll reference the system.dll library. This is the manual equivalent of adding a Web Reference in *Visual Studio.NET*.

many of the details. In this case, you need to tell the compiler to create a library, the name of the dll you create, and any references the proxy needs.

Now that you have a proxy, you need to put it in such a place that your consumers can see it. This is usually the bin directory under your wwwroot for *ASP.NET* pages. For example, on the author's machine the path is c:\Inetput\wwwroot\bin. For a C# executable, the reference is made while the executable is being compiled. Thus, it doesn't have to reside in any special place.

The code that comes at the top of the *ASP.NET* page is much simpler than that created by *Visual Studio.NET*. *Visual Studio* does a lot of code generation for each project, and we just don't need this here. All you have at the beginning of this page is a definition to explain that the code is written in C# and an import of the namespace created when the WSDL tool was used. Other than those two small details, the code is the same as our previous examples and executes the same.

```
<%@ Page language="c#" %>
<%@ Import Namespace="SDKStocks" %>
<html>
<script runat="server">
protected void Submit_Click(Object Src, EventArgs E)
{
XPlatformServices example = new XPlatformServices();
  string stockSymbol = SymbolToGet.Text;
  double value = 0;
  string stockValue = "";
  stockValue = example.GetTestQuote(stockSymbol).ToString();
lblMessage.Text = stockValue;
}
</script>
```

```
<body>
<form runat="server">
<b>Stock Symbol:</b>
    <asp:TextBox ID="SymbolToGet" Runat="server"/><br>

<b>Returned:</b>
    <asp:Label ID="lblMessage" Runat="server"/><br>
    <input type="submit"
            value="get quote"
            onserverclick="Submit_Click"
            runat="server"/>
</form>
</body>
</html>
```

Just by simply pointing your browser to this *ASP.NET* page on your system, the code compiles.

C# Client Compiled with *Framework SDK*

No example of a Web Services consumer would be complete without creating an executable client. C# makes it very easy to create such a client.

The following code is a simple command line tool to call the example Web Service. The beginning of the code calls the System namespace and then it defines the class XPlatformTest. The program uses the default constructor and goes right into defining the Main method. Two strings are defined to handle the value passed to the Web Service and then the result. Next the example object is created using the SDKStocks.XPlatformServices. Remember that SDKStocks is the namespace defined when you used the WSDL tool. The XPlatformServices is the class you defined in the Web Service code. Then the string myResult gets the result of the GetTestQuote method call. Note that this only works if the ToString()method is used as well. Finally, the myResult variable is written to the command line. The program simply echos 55.95 to the command line.

```
using System;
public class XPlatformTest
{
public static void Main()
{
String mySymbol = "C";
String myResult = "";
```

```
SDKStocks.XPlatformServices example = new
SDKStocks.XPlatformServices();
myResult = example.GetTestQuote(mySymbol).ToString();
Console.WriteLine(myResult);

    }
  }
```

To build the executable, you still have to reference the proxy dll created earlier. You just need to refer to it at compile time and the following example assumes that the dll is in the same directory. Notice that you only need to use the /r option to compile in the reference in order for the executable to use the information in the proxy.

```
csc /r:SDKStocks.dll XPlatformTest.cs
```

This creates XPlatformTest.exe, which when run from the command line, returns 55.95.

DISCOVERY AND *.NET*

An important part of Web Services is providing a means of discovery so that developers can easily find the functionality they need and may want to purchase. UDDI is part of the discovery process discussed in Chapter 5 and many features in Microsoft's *.NET* environment are an attempt in helping you with this process.

Disco

Disco is a command line tool that comes bundled with the *.NET Framework SDK*. If a site is set up properly, you can point the tool to a particular site and get a list of available Web Services. To enable this feature on your Web site that possesses Web Services, you need to modify your *index.html* file with the following information in the HEAD HTML tag, as shown in the following code.

```
<HEAD>
  <link type='text/xml'
        rel='alternate'
```

```
                    href='default.disco'/>
        </HEAD>
```

This small piece of meta information tells the disco command line tool where your discovery document (disco document) resides on your system. The rel attribute indicates that the disco document is available via a relative link and not on another Web site. The HREF directive indicates where the disco document resides on a particular system.

A disco file is simply an XML document that resides on your system and contains Web Services that you want others to discover in this manner.

Note that the disco XML format is a creation of Microsoft and not part of any of the standards discussed in earlier chapters.

Now go out to the DOS prompt in *Windows* and ensure that your *.NET* environment is set up properly by either already having the environment set up in DOS or by executing *corvars.bat*. Then you can execute the following command.

```
disco http://localhost/Default.disco /nosave
```

This reveals the available disco documents on the system. In this case, the response should look something like the following.

```
Microsoft (R) Web Services Discovery Utility
[Microsoft (R) .NET Framework, Version 1.0.3705.0]
Copyright (C) Microsoft Corporation 1998-2001. All
rights reserved.

Disco found documents at the following URLs:
http://localhost/Default.disco
```

Because the disco file is an XML document, you can use *Internet Explorer* to view the XML to reveal the available Web Services on a particular machine that utilizes disco. Figure 6.17 displays *Internet Explorer* viewing the disco file.

The XML file reveals to your user the location of a Web Service on your system.

With UDDI becoming more popular, the concept of Microsoft's disco tool becomes moot. UDDI provides a better interface to discovery and even provides you with the means to communicate with the repository through an API.

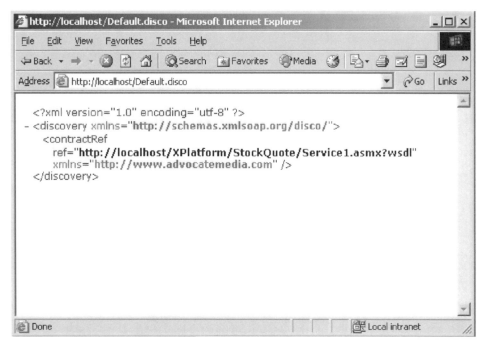

FIGURE 6.17   The disco file shown in *Internet Explorer*.

Web References, UDDI, and *Visual Studio.NET*

Within *Visual Studio.NET*, adding a Web reference provides you with an easy means of discovery. This tool integrates with *uddi.Microsoft.com* to help you find various Web Services. Utilizing it is a matter of clicking on the "Add Web Reference" selection in the "Project" menu of *Visual Studio.NET*.

By adding a Web reference, *Visual Studio.NET* creates the proxy for you and this is compiled with the entire project. Remember that with *SDK* the proxy had to be created manually.

In previous examples in this chapter, the "Add Web Reference" dialogue was used by adding a reference to a Web Service residing on the local machine. This tool is also useful for finding Web Services on the Web. Figure 6.18 shows the initial dialogue that appears. Notice on the lefthand side that you can search by service provider name.

Enter "Stock Quote" as the name of the service name. You will see results similar to those shown in Figure 6.19.

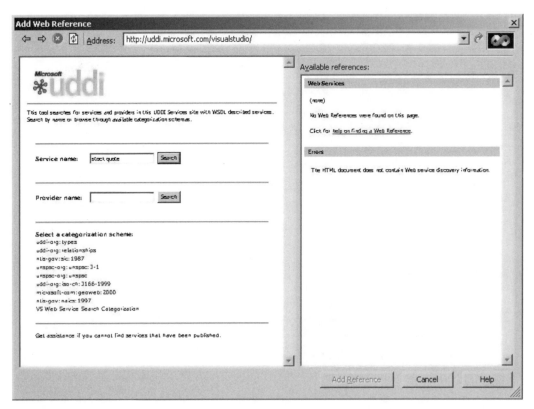

FIGURE 6.18 Searching for a Web Service using the "Add Web Reference" dialogue in *Visual Studio.NET*.

Now click on a link to the to the WSDL file in the listing for one of the stock quote Web Services. You'll see that the WSDL looks similar to what appears in Figure 6.20.

Then you can simply create the Web reference for your project.

Using UDDI Alone

Remember from Chapter 5 that you don't have do use the Microsoft tools to find the Web Service you need. There are several UDDI Web sites that offer UDDI services. So don't forget about using these directly.

WSDL Tool

Even though UDDI and WSDL share no formal relationship, WSDL is still part of the discovery process. The WSDL tool provides you with the ability to create

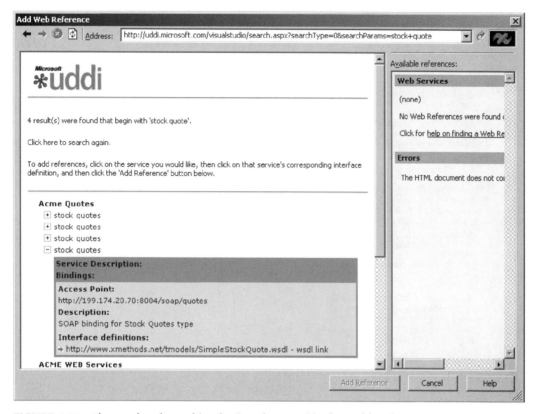

FIGURE 6.19   The results of searching for "Stock Quote" in the "Add Web Reference" dialogue.

proxies for your projects that are made with the *Framework SDK*. Remember that UDDI and WSDL do not share any formal relationship, but utilizing both standards is important for discovering the service and then discovering which methods are available.

The use of the WSDL tool is discussed in detail earlier in this chapter, when the creation of the proxy is discussed.

CONCLUSION

Microsoft's implementation of Web Services is probably easiest to use. Many of the more tedious details of using and creating Web Services, such as the generation of

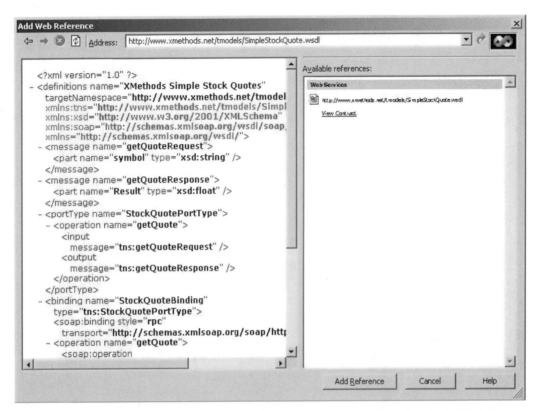

FIGURE 6.20 The WSDL file for one of the stock quote Web Services found in the "Add Web Reference" dialogue box.

WSDL, are done automatically for you. Plus, with the availability of the *.NET Framework SDK*, you can create and deploy these Web Services without a large budget.

The main problem with *.NET* is that it is available on only one platform. So if you have any UNIX platforms you need to integrate as part of your project, they will not be able to use *.NET*. You can get Web Services on UNIX to communicate with services on *.NET*; you just can't use the same pieces of software on each operating system.

As you move forward in the book, you'll see more complicated *.NET* examples and examples where different implementations of Web Services communicate.

7

Web Services with Apache *SOAP*

In This Chapter

- Prerequisites for Using Apache *SOAP*
- Creating and Deploying an Apache *SOAP* Web Service
- Creating Web Service Consumers in Java, Java Servlets, and JSP Pages

T he Apache Group provides a free *SOAP* library, Apache *SOAP* Version 2.3, to allow you to create and deploy Web Services across multiple platforms because it utilizes Java as the underlying language. This is an advantage if you have an environment containing a mix of operating systems because Java works on many platforms.

The main differences between Apache *SOAP* and the Web Services found in the *.NET* environment includes ease of use and installation. Microsoft's installation of either the *.NET Framework SDK* or *Visual Studio.NET* is seamless. You install it and you're ready to begin, and the code is at a very high level.

With Apache *SOAP*, the installation is far more complex. It is not difficult, but you must pay close attention to the installation instructions. If you miss one step or you don't have the environment set up right, you will find the installation of Apache *SOAP* to be somewhat frustrating.

The other difference is that coding for both creating a Web Service and calling a service happens at a lower level where you are more aware of the XML being used in the *SOAP* requests. Microsoft hides many of these implementation details from you.

This chapter covers how to create a simple Web Service with Apache SOAP and then create several different consumers ranging from command line clients to *Java Server Pages* (JSP). This chapter begins by covering the prerequisites your system needs to utilize Apache *SOAP*.

The examples in this chapter are based on using Apache SOAP Version 2.3. The latest release of Web Services software from the Apache Group is called Axis *and it is a complete rewrite of the Apache SOAP library. Both are currently available from the Apache Group's Web site at* xml.apache.org. *Chapter 8 provides an introduction to Apache* Axis *and explains how it differs from the SOAP library.*

PREREQUISITES FOR USING APACHE *SOAP*

There are several libraries and software components to install. The first piece to install is the Apache Group's Java container, *Tomcat*, which allows you to host the *SOAP* server in addition to servlets and JSPs. The next step after *Tomcat* involves installing all the necessary libraries including *SOAP*, *JavaMail*™, and others. Once the libraries are installed, the CLASSPATH environment variable needs to be installed correctly. For more information on where to download the various libraries and *Tomcat*, see Appendix B.

Installing *Tomcat*

On the Apache Group's Web site at *http://www.apache.org*, find the file named jakarta-tomcat-3.2.4.zip and unpack it to a directory on your system. This is a slightly older version of *Tomcat*, but it tends to be more stable and easier to work with. You should create a root directory for all the Apache software used in this chapter, and then install *Tomcat* under this directory. For example, on the author's machine all software used in this chapter is found under c:\xmlapache. This puts all the software close together and makes it easier to set CLASSPATH and other environmental variables.

Once *Tomcat* is extracted, you may want to change the name of the directory from c:\xmlapache\jakarta-tomcat-3-2-4 to c:xmlapache\tomcat.

This will save you some headache as you move forward in the chapter.

In this chapter, Tomcat *is used as a Web server and Java container. In a real production environment you need to use the Apache* Webserver *as your Web server.* Apache *directs calls when needed to* Tomcat *and is far more configurable. This gives your Web application and Web Services more options for security and redirects.*

If you don't already have Java installed on your system, download it from *www.javasoft.com* and then install it on your system. The examples in this chapter use Version 1.4.0 of the *Java Development Kit* (JDK).

Installing *SOAP* and Other Libraries

Now that Java is installed and *Tomcat* extracted, there are several other files you need to download. Appendix B describes the download files, the URL to download them from, and the recommended installation locations.

There are versions of the Apache Group's XML parser Xerces *that are newer than Version 1.4.4, but don't use them with this version of the Apache* SOAP *library. Otherwise you will receive various strange errors without much explanation from* Tomcat *and any client you compile.*

Once you have extracted all the zip files into the appropriate location for your system, you need to place the soap.war file into *Tomcat*'s webapps directory (e.g., webapps may have the following path): c:\xmlapache\tomcat\webapps). This is essentially a zip file that *Tomcat* extracts properly when it executes. This installs all the servlets and other code needed for the Apache *SOAP* library to execute.

Setting up the CLASSPATH

Now that all the libraries are installed on your machine, you need to tell Java where to find them. This is done by setting up the CLASSPATH environment variable correctly. Creating an MS-DOS batch file is one way to simplify the process. Using a batch file prevents your environment from becoming too polluted because you only set the environment variables that you are using at any particular time. Consider the following example bat file and remember that these examples assume you installed your Apache files in c:\xmlapache. Realize that the line breaks are put here to make it easier to read the code.

```
set PATH=%PATH%;c:\jdk1.4.0\bin;.
set JAVA_HOME=c:\jdk1.4.0
set CLASSPATH=c:\xmlapache\javamail\mail.jar;
              c:\xmlapache\jaf\activation.jar;
              c:\xmlapache\soap;
              c:\xmlapache\soap\lib\soap.jar;
              c:\xmlapache\tomcat\lib\servlet.jar;
              %CLASSPATH%;
```

ON THE CD
This batch file can be found on this book's CD-ROM in the code for this chapter and it's called env.bat. This batch file not only sets up the CLASSPATH but also the PATH so that we can execute Java and Javac, and JAVA_HOME is set in order for *Tomcat* to find the Java root directory.

When the CLASSPATH is set, note that it calls the Java Archive Files (JAR) that you expected but it also includes servlet.jar in the *Tomcat* directory structure. The *SOAP* library needs this to execute the servlets that come with the distribution.Examples later in this chapter also utilize servlets.

Finally, to get Tomcat to load the correct XML parser for *SOAP*, you need to change one of the lines in its startup script. Load the startup.bat batch file into a text editor and change the line that sets the CP environment variable (this is the CLASSPATH for *Tomcat*) to look like the following.

```
set CP=path-to-xerces\xerces.jar;%CLASSPATH%;%CP%
```

Testing the Installation of Apache *SOAP*

To begin testing the Apache *SOAP* installation, start *Tomcat* with the startup script found in its bin directory. Make sure you use the env.bat script to set up the environment at a DOS prompt and then change into *Tomcat*'s bin directory to start the container.

To test your installation, point your browser to the following URL: *http://localhost:8080/soap/*. You should see what's shown in Figure 7.1.

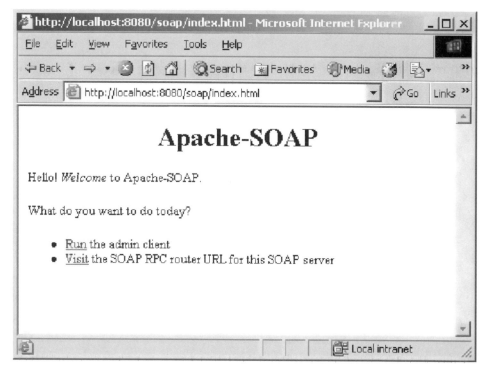

FIGURE 7.1   The appropriate response to a browser after installing the Apache *SOAP* library.

This means that *Tomcat* successfully extracted the *SOAP* war file. Click on the link for the "Admin" tool and you should see what appears in Figure 7.2.

If you visit the *SOAP* RPC router, it doesn't really respond to the browser but the results still look like Figure 7.3.

FIGURE 7.2 How the *SOAP* "Admin" tool appears in *Internet Explorer*.

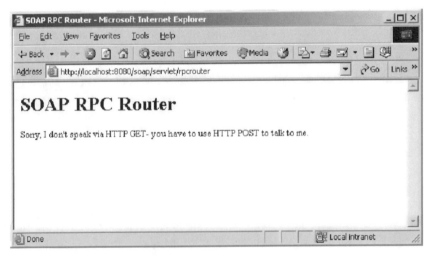

FIGURE 7.3 The *SOAP* RPC router in *Internet Explorer*.

These different Web pages primarily show that the war file installed properly but do not test the functionality. The *SOAP* distribution comes with several examples that can help you test the installation. In `c:\xmlapache\soap\samples` you will find a MSDOS batch file named `testit.cmd`. This will install, test, remove a Web Service that returns the value of IBM's stock price, and then exit. If it works properly you see the results in a DOS prompt similar to Figure 7.4.

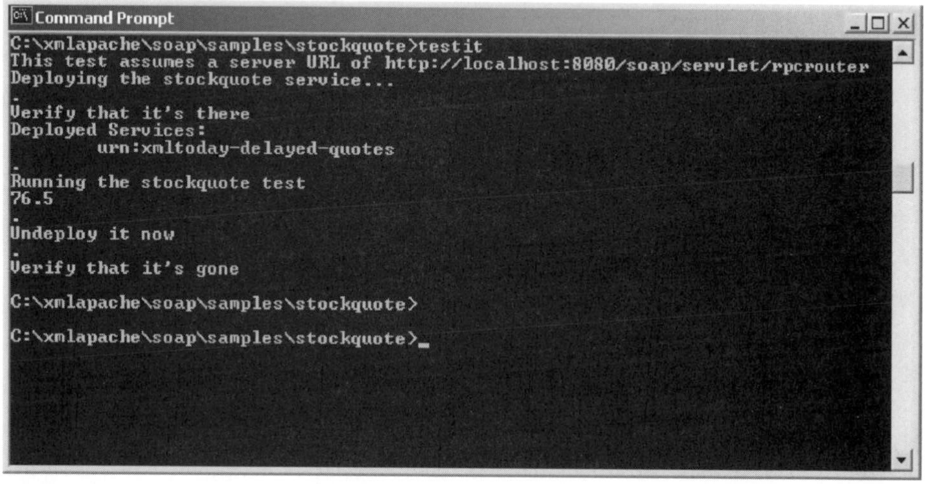

FIGURE 7.4   The results of running `testit.cmd` in the Stock Quote sample directory of Apache *SOAP*.

If you receive an error like the following when you execute `testit.cmd`, the `CLASSPATH` for *Tomcat* is not set up properly.

```
java org.apache.soap.server.ServiceManagerC
client http://localhost:8080/soap/servlet/rpcrouter list
Exception in thread "main" [SOAPException:
faultCode=SOAP-ENV:Protocol; msg=Unsupported
response content type &quot;text/html&quot;, must be:
&quot;text/xml&quot;. Response was:

&lt;h1&gt;Error: 500&lt;/h1&gt;
&lt;h2&gt;Location:
/soap/servlet/rpcrouter&lt;/h2&gt;&lt;b&gt;Internal
ServletError:&lt;/b&gt;&lt;br&gt;&lt;pre&gt;
java.lang.NoClassDefFoundError: javax/mail/
```

Now that you've seen how to test Apache *SOAP*, the next sections describe how to create Apache *SOAP* Web Services and their consumers.

CREATING AND DEPLOYING AN APACHE *SOAP* WEB SERVICE

The Apache *SOAP* implementation is powerful because Java gives you a lot of options for utilizing Web Services because you have the ability to compile code on multiple platforms with Java. Microsoft *.NET*, on the other hand, is limited for now to just *Windows*. Thus, the Apache *SOAP* library offers a great deal more flexibility. There are, however, rumors of *.NET* being ported to Linux.

The code needed to develop a Web Service appears similar to the code used to generate a Web Service in Chapter 6, in C#. In some ways it is simpler because *Visual Studio.NET* generates so much code, which works behind the scenes, for every Web Service.

This section covers writing and compiling the code for an Apache Web Service and the process for deploying it.

The Web Service

The following code is the Java version of our `SimpleStockQuote` example that has been used throughout this book. It starts out with the package declaration. Note that the statement `package samples.simplestock` indicates that this example exists in a directory structure `/samples/simplestock` somewhere in your CLASSPATH. To compile this example, you'll need the environment on your system to match the one discussed in the prerequisites section of this chapter.

The first two import statements are standard libraries for handling URLs and Java input and output. The second two import statements, which include `org.w3c.dom.*` and `org.xml.sax.*` bring in functionality for different models of parsing XML. The `javax.xml.parsers.*` allows a program to obtain and manipulate an XML parser, and finally `org.apache.soap.util.xml.*` allows access to the functionality in the Apache *SOAP* library. You don't really need to worry about which import statements are used because these examples use the imports that all Apache Web Service code needs.

After all the import statements, the code finally gets to the functionality that allows the Web Service to work. First, the class `SimpleStockQuote` is defined. Note that the file must be named `SimpleStockQuote.java` to compile it. (This is different from

C#.) Then there is the simple code for the method `getTestQuote`, which returns the same values as all the other Stock Quote examples in this book.

```
package samples.simplestock;

import java.net.URL;
import java.io.*;
import org.w3c.dom.*;
import org.xml.sax.*;
import javax.xml.parsers.*;
import org.apache.soap.util.xml.*;

public class SimpleStockQuote {

 public String getTestQuote (String symbol) throws Exception {

   if (symbol.equals("C") )  {
      return "55.95";
   } else {
      return "-1";
   }


 }
```

Now that the service is written, you can compile it by typing `javac SimpleStockQuote.java`. This will create the `SimpleStockQuote.class` file which is the compiled byte code Apache *SOAP* needs to execute the Web Service.

In order for you to get this Web Service working quickly, put this Web Service in the samples directory of your Apache SOAP installation.

NOTE

Deploying the Service

To deploy an Apache *SOAP* Web Service, creating an XML file called a "deployment descriptor" is necessary to register a service with the server. It identifies the class, methods, and the name used to expose the functionality via Web Services.

The following XML example is the deployment descriptor for the Web Service created in the previous example. The root element defines the needed namespace and the ID attribute describes the name needed to call the methods in that service.

When you write the client for this example, the URN called will be `simple-stock-quote`. The provider element defines the exposed method and the type of Web Service. In this case, you used Java. The scope attribute has three possible values: request, session, and application. Table 7.2 explains each option.

TABLE 7.2 The Three Options for the Scope Attribute in the Deployment Descriptor

Option	Purpose
Request	The *SOAP* implementation removes the object once the request is complete.
Session	The values a particular object holds only last the life of the HTTP session.
Application	The object lasts as long as the servlet that is managing the *SOAP* object is in service.

The `isd:java` tag describes the location of the class. Remember that `Simple-StockQuote` was part of the `samples.simplestock` package.

Finally, the fault listener tag describes the class that handles any *SOAP* fault returned from the server. In this case, it is the default class provided by the Apache group. If you needed to, you could add your own fault listener to provide custom error handling for an application.

```
<isd:service xmlns:isd="http://xml.apache.org/xml-soap/deployment"
             id="urn:simple-stock-quote">
  <isd:provider type="java"
                scope="Application"
                methods="getTestQuote">
      <isd:java class="samples.simplestock.SimpleStockQuote"/>
  </isd:provider>

<isd:faultListener>
    org.apache.soap.server.DOMFaultListener
</isd:faultListener>
</isd:service>
```

Now that you have a deployment descriptor, you must use tools provided with the Apache *SOAP* library to actually deploy the Web Service.

Deploying SimpleStockQuote

Now that you have a deployment descriptor, you can execute the following class to complete deployment.

```
org.apache.soap.server.ServiceManagerClient.
```

Ensure that you have the CLASSPATH set up as described earlier in the chapter.

If you are unable to execute the following examples, not having the CLASSPATH configured properly is the most likely culprit. The second most likely cause is not having your path set up so you can execute Java. Also check to ensure that you modified the startup script of Tomcat.

Start *Tomcat* and allow it to load all the different webapps that are installed. This includes /soap. To install the SimpleStockQuote example, execute the following from the directory where you have the code.

```
java org.apache.soap.server.ServiceManagerClient
http://localhost:8080/soap/servlet/rpcrouter deploy
DeploymentDescriptor.xml
```

If you receive a connection refused error, you probably haven't started Tomcat.

When you execute the deploy command, no information comes back to let you know that the deployment was successful other than that there was no error. Looking in the *SOAP* Admin tools will confirm the installation of SimpleStockQuote. By going to *http://localhost:8080/soap/admin/index.html* you can see what functionality has been deployed. Click on the "List" icon and you should see the SimpleStockQuote Web Service as the only deployed Web Service. Figure 7.5 shows the results in *Internet Explorer*. Remember that in the deployment descriptor you gave the Web Service the following unique identifier: urn:simple-stock-quote. This is how the Admin tool lists the various services.

Another option for listing the available services on a particular system is to execute the following command from the DOS prompt.

FIGURE 7.5 How the *SOAP* Admin tool lists the `SimpleStockQuote` Web Service.

```
java org.apache.soap.server.ServiceManagerClient
http://localhost:8080/soap/servlet/rpcrouter list
```

The response is a list of deployed services and yours should simply look like this:

```
Deployed Services:
    urn:simple-stock-quote
```

The next section covers how to secure the deployment of Web Services to your server.

Securing the Deployment of Web Services

With the default configuration of Apache *SOAP*, anyone can deploy Web Services to your system, and that's very dangerous from a security perspective. This is fine when the server is in a development environment where it is protected by a firewall and possibly a subnet, but when you put your system into production you want to be sure that you control who installs Web Services.

By changing one configuration file and adding another, you can turn the remote installation of Web Services off. The first step is to modify the *SOAP* distribution's web.xml file. This is the configuration file for the servlets that support the RPC router and the "Admin" tool. If you installed *Tomcat* into the suggested directory, the web.xml file resides in the c:\xmlapache\tomcat\web-apps\soap\WEB-INF directory. When opened, this file should look like the following.

```
<web-app>
  <display-name>Apache-SOAP</display-name>
  <description>no description</description>

  <servlet>
    <servlet-name>rpcrouter</servlet-name>
    <display-name>Apache-SOAP RPC Router</display-name>
    <description>no description</description>
    <servlet-class>
        org.apache.soap.server.http.RPCRouterServlet
    </servlet-class>
    <init-param>
... snip snip
```

This is just a sample of first part of the file because you only need to modify the beginning. Right after the description tag, add the following information.

```
<context-param>
    <param-name>ConfigFile</param-name>
    <param-value>c:/xmlapache/soap.xml</param-value>
</context-param>
```

This creates a pointer to a configuration file, soap.xml, where you can tell Apache *SOAP* to do different things. In the soap.xml file, add the following XML code.

```
<soapServer>
  <serviceManager>
  <option name="SOAPInterfaceEnabled" value="false" />
  </serviceManager>
</soapServer>
```

This disables your ability to add Web Services from the command line. If you try to add a service, you will get the following response.

```
Ouch, the call failed:
    Fault Code    = SOAP-ENV:Server.BadTargetObjectURI
    Fault String = Unable to determine object id from
    call: is the method element namespaced?
```

Even though you can now prevent people from installing Web Services from a DOS or other command prompt, users coming from the Web can still access the "Admin" tool. That's why it's important to use Tomcat *with a Web server like* Apache. *Apache has ability to protect certain Web pages from being accessed and even also provides a means of authorization.*

Viewing the *SOAP* Messages

In Microsoft's implementation of *.NET* Web Services, the *SOAP* messages are part of the Web page the service generates. With Apache *SOAP*, there is a means of looking at the messages of Apache *SOAP* and other implementations using the *TCPTunnelGui* tool. This tool provides you with a means of viewing the actual message the client and server are exchanging. To execute the tool, use the following command from a DOS prompt. (Note the returns are added to make it easier for you to read.)

```
java org.apache.soap.util.net.TcpTunnelGui 8000
    localhost 8080
```

The *TCPTunnelGui* acts as a proxy to the Web Service requests and response. It takes the request, prints it to the screen, and then forwards it to the server. It does the same thing with the response. Thus, the number "8000" in the previous example is the port it is listening to, `localhost` is the server it forwards requests to, and 8080 is the port that server is listening to. Then you must modify your client to send requests to port 8000 to make this work. Figure 7.6 illustrates *TCPTunnelGui* showing the request and response messages for the `SimpleStockQuote` example.

For closer examination, here is the complete *SOAP* request for the `Simple-StockQuote` example.

```
Content-Type: text/xml;
charset=utf-8 Content-Length: 461
SOAPAction: ""
<?xml version='1.0' encoding='UTF-8'?>
```

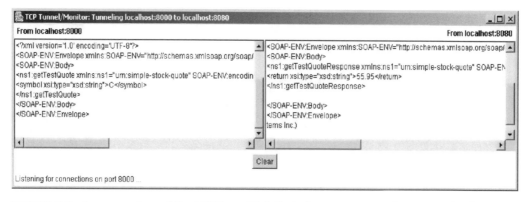

FIGURE 7.6 Screen capture of the *TCPTunnelGui* displaying a request and response for the `SimpleStockQuote` example.

```
<SOAP-ENV:Envelope xmlns:SOAP-
        ENV="http://schemas.xmlsoap.org/soap/envelope/"
            xmlns:xsi="http://www.w3.org/2001/XMLSchema-instance"
            xmlns:xsd="http://www.w3.org/2001/XMLSchema">
   <SOAP-ENV:Body>
     <ns1:getTestQuote xmlns:ns1="urn:simple-stock-quote"
            SOAP-
ENV:encodingStyle="http://schemas.xmlsoap.org/soap/encoding/">
            <symbol xsi:type="xsd:string">C</symbol>
     </ns1:getTestQuote>
   </SOAP-ENV:Body>
   </SOAP-ENV:Envelope>
```

<t>And the response from the Server for SimpleStockQuote:

```
      HTTP/1.0 200 OK
Content-Type: text/xml;
charset=utf-8 Content-Length: 483
Set-Cookie2: JSESSIONID=3bymuj1261;Version=1;Discard;Path="/soap"
Set-Cookie: JSESSIONID=3bymuj1261;Path=/soap
Date: Thu, 08 Aug 2002 00:45:49 GMT
Servlet-Engine: Tomcat Web Server/3.2.4 (JSP 1.1; Servlet 2.2; Java
      1.4.0;
Windows 2000 5.0 x86; java.vendor=Sun Microsys

   <?xml version='1.0' encoding='UTF-8'?>
   <SOAP-ENV:Envelope xmlns:SOAP-
```

```
          ENV="http://schemas.xmlsoap.org/soap/envelope/"

          xmlns:xsi="http://www.w3.org/2001/XMLSchema-instance"

          xmlns:xsd="http://www.w3.org/2001/XMLSchema">

    <SOAP-ENV:Body>
          <ns1:getTestQuoteResponse xmlns:ns1="urn:simple-stock-quote"
          SOAP-
ENV:encodingStyle="http://schemas.xmlsoap.org/soap/encoding/">
          <return xsi:type="xsd:string">55.95</return>
          </ns1:getTestQuoteResponse>
    </SOAP-ENV:Body>
    </SOAP-ENV:Envelope>
```

These request and response messages look similar to the other *SOAP* examples shown in this book so far.

Now that you understand how to create, deploy, and monitor an Apache *SOAP* Web Service, you should be ready to create consumers in Java.

CREATING WEB SERVICE CONSUMERS IN JAVA, JAVA SERVLETS, AND JSP PAGES

A consumer, as mentioned in previous chapters, is a piece of software or a Web page that actually utilizes a Web Service. In this chapter, the consumers will be a command line Java application, a servlet, and a *Java Server Page* (JSP).

Command Line Application

The command line application in this chapter is similar to the first C# application shown in Chapter 6. The first three import statements are just general Java functionality needed to read and write files along with handling URLs.

NOTE

Unlike C#, the file that you compile for this small application must be named after the class. In this case, the class is GetQuote *so the file must be called* GetQuote.java.

All the functionality in this example occurs in the main method. The code starts off by defining the *SOAP* encoding style and the URL to call the Web Service. It then puts the one command line argument into the string symbol.

Then the actual call to the Web Service is created. The particular Web Service called is defined as `urn:simple-stock-quote`. Remember that this is the unique identifier created in the deployment descriptor shown earlier in the chapter.

The vector params is the structure into which the value of symbol is placed. Once in the vector, the `setParams` method adds params to the call. Next, the response object receives the values returned by the invoke method of the call object. Finally, there is some error handling to catch errors generated by the *SOAP* request. If there is no error, the following snippet displays the result.

```
Parameter result = resp.getReturnValue ();
System.out.println (result.getValue ());
```

The complete code example:

```
import java.io.*;
import java.net.*;
import java.util.*;
import org.apache.soap.*;
import org.apache.soap.rpc.*;

public class GetQuote {
  public static void main (String[] args)
  throws Exception {

  String encodingStyleURI =  Constants.NS_URI_SOAP_ENC;
    //specify the URL of the Web Service
    URL url =new URL
  ("http://localhost:8080/soap/servlet/rpcrouter");
    //Get the string passed in at the command line.
    String symbol = args[0];
    System.out.print("****" + symbol + "****");

    //Create the call to get a specific Web Service
    Call call = new Call ();
    call.setTargetObjectURI ("urn:simple-stock-quote");
    call.setMethodName ("getTestQuote");
    call.setEncodingStyleURI(encodingStyleURI);

    //Create the vector that has the string value.
    Vector params = new Vector ();
    //Add the string to the vector
    params.addElement (new Parameter("symbol",
    String.class, symbol, null));
```

```
                call.setParams (params);
                //grab the response
                Response resp = call.invoke ( url,"" );

                //if there's a fault, catch it.
                //else display the result
                if (resp.generatedFault ()) {
                  Fault fault = resp.getFault ();

                  System.err.println("Generated fault: " + fault);
                } else {
                  Parameter result = resp.getReturnValue ();
                  System.out.println (result.getValue ());
                }
              }
            }
```

The next section takes this example one step further by taking the code and making a GUI application using Java *Swing*™.

Java *Swing* Example

Creating a GUI that calls the Web Service in Java is just slightly more work that the command line application shown in the previous example. You need to create the different components that make up the GUI, and then use an event listener to listen for actions happening within the application.

The first step is to import all the proper libraries. In this case, you need to import not only the classes for the Apache *SOAP* but also for *Swing* and *AWT*™. *Swing* and *AWT* are the libraries within Java that allow you to create *Windows* applications that are cross platform. The *AWT* library is more operating-system-dependent than *Swing*, but it still supports many of the *Swing* objects. The rest of the imports are used in the previous examples in this chapter.

The next step is to define the class called WebSvcGui. Notice the keywords extends and implements in the class definition. Extends indicates that the class inherits either all or some of the Frame class whereas implements indicates that the class adheres to the interface defined in ActionListener.

After the class definition, the constructor for the class creates the GUI. Notice the definitions for all the frames and buttons used in the GUI. There are two JTextField definitions and this is where values are read from and written to. The action listener calls the Web Service's static method callService(). The action lis-

tener waits for something to happen to one of the objects in the GUI, which is the
`valueButton` object. The `callService()` method is static so you don't need to instantiate another copy of the object to call the method from the constructor.

Notice that the definition for the `callService` method is doing more error handling than the command line application. The `Swing` classes require this extra error
handling so the GUI knows how to react. Other than the handling this, the code to
call the Web Service isn't different than the previous example.

Then, main method creates an instance of `WebSvcGui` and makes it visible. Remember that all actions that occur in the GUI are taken care of by an action
listnener.

```java
import java.awt.*;
import java.awt.event.*;
import javax.swing.*;
import java.io.*;
import java.net.*;
import java.util.*;
import org.apache.soap.*;
import org.apache.soap.rpc.*;


public class WebSvcGui extends Frame implements
ActionListener {

public WebSvcGui () {
  //name the window
  super("Web Service Gui Example");
  setSize(400,116);

  //create panel and size
  JPanel myPanel = new JPanel();
  myPanel.setLayout(new GridLayout(3,2,5,5));

  JLabel symbol = new JLabel("Symbol:", JLabel.RIGHT);
  final JTextField symbolField = new JTextField(10);

  JLabel result = new JLabel("Result", JLabel.RIGHT);
  final JTextField resultField = new JTextField(10);

  //add a button
  JButton valueButton = new JButton("Get Value");
```

```
        //create a listener for the button. It's listening
        //for the button to be clicked.
        //Call the Web Service from the listener
        valueButton.addActionListener(new ActionListener() {
        public void actionPerformed(ActionEvent ev) {
          String symbol = symbolField.getText();
          //call the static Web Service method
          String returned = WebSvcGui.callService(symbol);
          resultField.setText(returned);
        }
      });

    myPanel.add(symbol);
    myPanel.add(symbolField);
    myPanel.add(result);
    myPanel.add(resultField);

    myPanel.add(valueButton);
    //add the panel to the gui.
    add(myPanel, BorderLayout.SOUTH);
  }

  public void actionPerformed(ActionEvent ae) {
    System.out.println(ae.getActionCommand());
  }
  //static so we can call it without creating an WebSvcGui
  //object
  static String callService(String symbol) {

    String encodingStyleURI =  Constants.NS_URI_SOAP_ENC;
    URL url = null;
    //set up URL to call
    try {
     url = new URL
          ("http://homer:8080/soap/servlet/rpcrouter");
    } catch (MalformedURLException e){
      System.out.println("Error:" + e);
    }

    //set up to the call to the Web Service
    Call call = new Call ();
    call.setTargetObjectURI ("urn:simple-stock-quote");
```

```
call.setMethodName ("getTestQuote");
call.setEncodingStyleURI(encodingStyleURI);
Vector params = new Vector ();
params.addElement (new Parameter("symbol",
String.class, symbol, null));
call.setParams (params);
Response resp;

//try to call the Web Service, if successful return the
//value
try {
  resp = call.invoke ( url, "" );
  if (resp.generatedFault ()) {
     Fault fault = resp.getFault ();
     System.out.println("fault: " + fault );
     return("-1");
  } else {
     Parameter result = resp.getReturnValue ();
     return(result.getValue().toString());
  }


} catch (SOAPException e) {
   System.out.println("Error:" + e);
}

}

public static void main(String args[]) {
  WebSvcGui myGui = new WebSvcGui();
  myGui.setVisible(true);


}
}
```

Figure 7.7 reveals how the GUI appears running under *Windows 2000*.

The previous two examples focus on creating Java applications with Web Services. In the next couple of sections, the examples focus on consumers that communicate with a Web browser.

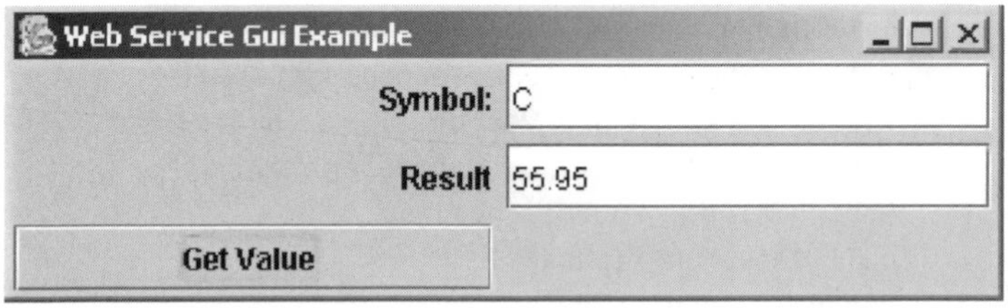

![Web Service Gui Example window showing Symbol: C, Result: 55.95, and a Get Value button]

FIGURE 7.7 How the `WebSvcGui` example appears under *Windows 2000*.

Servlet Consumer

A servlet is a chunk of Java code that executes on the server side for a request from a Web browser in a container such as *Tomcat*. A servlet implementation routes requests to the various Web Services within Apache *SOAP*.

The main difference between the following example and the previous is that a servlet has no main method. It executes as an extension of `HttpServlet` which the `extends` statement defines. In addition, there are more `include` files because you need all the functionality that any other servlet would normally have.

After the definition of the class `SimpleStockClient`, two methods must be defined: `doGet` and `doPost`. These are necessary for the servlet to understand requests from a browser.

A browser can send two different types of requests to a Web site. The Post *method is one way a browser sends information, usually through a form in the body of the submission. The other is a* Get *and this occurs when a URL contains data in the query string.*

In this example, the `doGet` method contains all the functionality, and then the `doPost` method just routes the request to `doGet`, as the following code snippet shows.

```
public void doPost(HttpServletRequest request,
                   HttpServletResponse response)
   throws IOException, ServletException
   {
       doGet(request, response);
   }
```

Modifying this example to do different things dependent on the type of request simply involves modifying the `doPost` method.

This example must handle HTML because the response from this example is sent to the browser, and you can see there are several `println` statements sending HTML back to the browser. This makes the servlet code hard to read and this is one of the reasons JSP was created. There are some JSP examples later in the chapter.

```java
import java.io.*;
import java.net.*;
import java.text.*;
import java.util.*;
import org.apache.soap.*;
import org.apache.soap.rpc.*;
import javax.servlet.*;
import javax.servlet.http.*;

//define the servlet
public class SimpleStockClient extends HttpServlet {

    //handle a get request
    public void doGet(HttpServletRequest request,
                      HttpServletResponse response)
        throws IOException, ServletException
    {
        //start the HTML response
        response.setContentType("text/html");
        PrintWriter out = response.getWriter();

        out.println("<html>");
        out.println("<head>");

        String title = "Get Simple Stock Quote";

        out.println("<title>" + title + "</title>");
        out.println("</head>");
        out.println("<body bgcolor=\"white\">");
        out.println("<body>");
        out.println("<h1>" + title + "</h1>");

        String encodingStyleURI =
        Constants.NS_URI_SOAP_ENC;
```

```
//the java code should now resemble the
//small app shown previously
URL url =new URL
("http://homer:8080/soap/servlet/rpcrouter");
String symbol = "C";
out.println("****" + symbol + "****");


Call call = new Call ();
call.setTargetObjectURI ("urn:simple-stock-quote");
call.setMethodName ("getTestQuote");
call.setEncodingStyleURI(encodingStyleURI);
Vector params = new Vector ();
params.addElement (new Parameter("symbol",
String.class, symbol, null));
call.setParams (params);
Response resp;
//catch any errors.
try {
   resp = call.invoke ( url, "" );
       if (resp.generatedFault ()) {
           Fault fault = resp.getFault ();
           out.println("<h2><font color=\"red\">
               Generated fault: " + fault + "
               </font></h2>");
       } else {
           Parameter result = resp.getReturnValue ();
           out.print("Result of Web Service call: ");
           out.println (result.getValue ());
       }
} catch ( SOAPException e ) {
   out.println(" unable to call Web Service " + e);
}

out.println("</body>");
out.println("</html>");
   }
}
```

Figure 7.8 displays the output of the first example servlet in *Internet Explorer*. The next example is a little more functional because it allows you to enter a value and submit it to the server to get a response. This is just a matter of adding

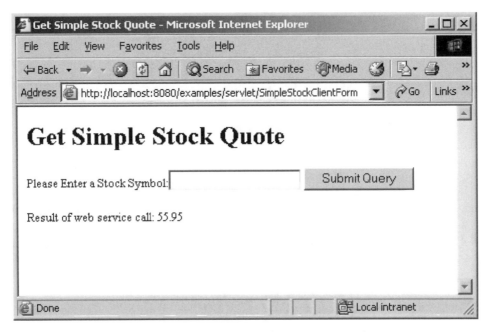

FIGURE 7.8 The output of the first example servlet in *Internet Explorer*.

some more `println` statements that have the appropriate HTML code for a form. Then an if statement tests to see if `symbol` has a value. If it does, it calls the appropriate Web Service with the value for `symbol`.

```
import java.io.*;
import java.net.*;
import java.text.*;
import java.util.*;
import org.apache.soap.*;
import org.apache.soap.rpc.*;
import javax.servlet.*;
import javax.servlet.http.*;

public class SimpleStockClientForm extends HttpServlet {


    public void doGet(HttpServletRequest request,
                      HttpServletResponse response)
        throws IOException, ServletException
```

```
{
    response.setContentType("text/html");
    PrintWriter out = response.getWriter();

    out.println("<html>");
    out.println("<head>");

    String title = "Get Simple Stock Quote";

    out.println("<title>" + title + "</title>");
    out.println("</head>");
    out.println("<body bgcolor=\"white\">");
    out.println("<body>");
    out.println("<h1>" + title + "</h1>");
    out.println("<P>");

    //This begins the definition of the form so
    //that you can send data to the servlet.
    out.println("<form
    action=\"./SimpleStockClientForm\" method=POST>");

    out.print("Please Enter a Stock Symbol:");
    out.println("<input type=text
                        size=20
                        name=symbol>");
    out.println("<input type=submit>");
    out.println("</form>");


    //most of the code is similar to the last
    //servlet example.
    String encodingStyleURI =
    Constants.NS_URI_SOAP_ENC;

    URL url =new URL
    ("http://homer:8080/soap/servlet/rpcrouter");
    String symbol = request.getParameter("symbol");

//if a value has been sent by the form go ahead
//process the data.
if(symbol != null && symbol.length() != 0) {

    Call call = new Call ();
```

```
call.setTargetObjectURI ("urn:simple-stock-quote");
call.setMethodName ("getTestQuote");
call.setEncodingStyleURI(encodingStyleURI);
Vector params = new Vector ();
params.addElement (new Parameter("symbol",
String.class, symbol, null));
call.setParams (params);
Response resp;

try {
    resp = call.invoke (/* router URL */ url, /* actionURI */
      "" );
     if (resp.generatedFault ()) {
       Fault fault = resp.getFault ();
       out.println("<h2><font color=\"red\">
       Generated fault: " + fault + "
       </font></h2>");
     } else {
        Parameter result = resp.getReturnValue ();
        out.print("Result of Web Service call: ");
        out.println (result.getValue ());
      }
  } catch ( SOAPException e ) {
      out.println(" unable to call Web Service " + e);
  }
 }

out.println("</body>");
out.println("</html>");
}

public void doPost(HttpServletRequest request,
                  HttpServletResponse response)
throws IOException, ServletException
{
    doGet(request, response);
}


}
```

By adding the form functionality, this example becomes more realistic. Figure 7.9 shows this example in *Internet Explorer*.

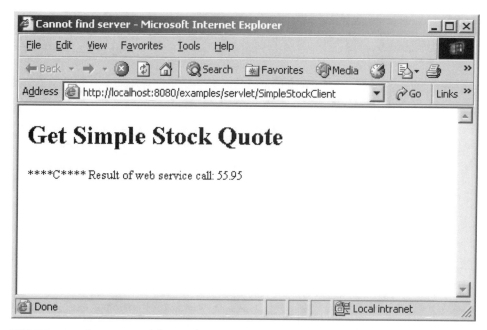

FIGURE 7.9 The output of the servlet form example after entering "C" in the text box.

Servlets are very useful for delivering the output of complex Java code to a browser, but if you have a lot of HTML to deliver the coding becomes difficult. JSP pages allow you to separate coding from content, and thus may provide you a little more flexibility as a Web Services consumer.

Using JSP to Code a Consumer

JSP simplifies server-side Java coding because it allows you to develop separate complex Java code from the actual display of information. Servlets display HTML with a series of `println` statements, but JSP pages mix the HTML with the Java code.

It is interesting to note that the first time JSP code executes the Java server, such as Apache Tomcat, actually generates servlet code from the template the JSP page defined. The servlet code then compiles and executes, and it sends the response back to the browser.

JSP can use Java code, but it also implements many HTML-like tags to perform many of the same functions as the Java code. Take a look at the first JSP consumer and compare it to one of the servlet examples, and you'll find that they are very different.

Simple JSP Consumer

The following code example shows how different JSP code is. The import statements are replaced with `<%@ page import .. %>`. Instead of using `println` statements to display HTML, the HTML mixes in with the JSP elements. Then the Java code that calls the Web Service is the same code used in the previous examples. The difference here is that the code appears between the <% %> tags. Code that appears between these tags are referred to as *scriptlets*.

```
<%@ page import = "java.io.*, java.net.*,
    java.util.*, org.apache.soap.*,
    org.apache.soap.rpc.*, java.util.*" %>

<HTML>
  <HEAD>
    <TITLE>JSP Web Services Page</TITLE>
  </HEAD>
  <BODY>
  <H2>JSP Web Services Page</H2>

  <%
  String encodingStyleURI =  Constants.NS_URI_SOAP_ENC;
  URL url =new URL
  ("http://homer:8080/soap/servlet/rpcrouter");
  String symbol = "C";

   Call call = new Call ();
   call.setTargetObjectURI ("urn:simple-stock-quote");
   call.setMethodName ("getTestQuote");
   call.setEncodingStyleURI(encodingStyleURI);
   Vector params = new Vector ();
   params.addElement (new Parameter("symbol",
   String.class, symbol, null));
   call.setParams (params);
   Response resp;

   try {
```

```
            resp = call.invoke ( url,  "" );
            if (resp.generatedFault ()) {
                Fault fault = resp.getFault ();
                out.println("<h2><font color=\"red\">Generated
                fault: " + fault + " </font></h2>");
        } else {
                Parameter result = resp.getReturnValue ();
                out.print("Result of Web Service call: ");
                out.println (result.getValue ());
        }
        } catch ( SOAPException e ) {
            out.println(" unable to call Web Service " + e);
        }
    %>
        </BODY>
    </HTML>
```

Figure 7.10 shows *Internet Explorer* viewing the JSP response.

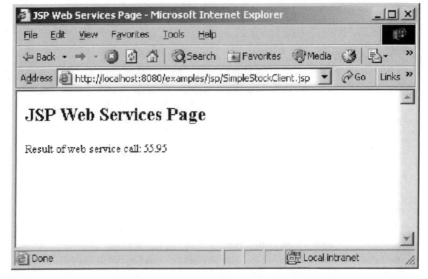

FIGURE 7.10 The first JSP example displaying the value from the Web Service.

JSP with Include Directive

The goal of JSP is to separate content from functionality. In the last example, much of the Java code still mixes in with the JSP and HTML elements. One way around having all the Java within the scriplet tags <% and %> is to use a jsp:include directive and have all the Java code appear in the separate file.

Consider the following simple JSP page.

```
<HTML>
  <HEAD>
    <TITLE>JSP Include Example</TITLE>
  </HEAD>

<H2>JSP Include Example</H2>
<jsp:include page="IncludeExample.jsp" flush="true"/>

</HTML>
```

It simply includes a JSP page, which contains the following code.

```
<%@ page import = "java.io.*, java.net.*, java.util.*,
  org.apache.soap.*, org.apache.soap.rpc.*,
  java.util.*" %>

<%
  String encodingStyleURI = Constants.NS_URI_SOAP_ENC;
  URL url =new URL
  ("http://homer:8080/soap/servlet/rpcrouter");
  String symbol = "C";

  Call call = new Call ();
  call.setTargetObjectURI ("urn:simple-stock-quote");
  call.setMethodName ("getTestQuote");
  call.setEncodingStyleURI(encodingStyleURI);
  Vector params = new Vector ();
  params.addElement (new Parameter("symbol",
                     String.class, symbol, null));
  call.setParams (params);
  Response resp;

  try {
    resp = call.invoke (url, "" );

    if (resp.generatedFault ()) {
      Fault fault = resp.getFault ();
      out.println("<h2><font color=\"red\">Generated
      fault: " + fault + " </font></h2>");
```

```
      } else {
         Parameter result = resp.getReturnValue ();
         out.print("Result of Web Service call: ");
         out.println (result.getValue ());
      }
   } catch ( SOAPException e ) {
      out.println(" unable to call Web Service " + e);
   }
```

This is the same code that appears in the first JSP example, but it appears in a separate JSP file. The output in *Internet Explorer* still appears the same.

Beans and JSP

Another way of separating complex Java code from the internal workings of a JSP is to put that code in a Bean. A Bean is a library of Java code that is similar to a dll in Microsoft implementations. It is accessible to JSP pages via *Tomcat*, and by using set and get methods the Bean and JSP page are able to communicate.

NOTE

Tomcat *looks for Beans in the classes subdirectory of* WEB-INF. *Within that directory you need to create a subdirectory and place your Bean within it. The name of that directory (or an entire path from the class's directory) needs to be defined with a package directive at the beginning of the Bean. If it isn't in a subdirectory, Tomcat attempts to load the class file as a servlet.*

By using Beans, the JSP pages you implement become very clean because all the complex Java code is hidden. In the following JSP example, there is not scriptlet code. All the communications with Java is done with the jsp:usebean, jsp:-setProperty, and the jsp:getProperty tags.

The jsp:usebean tag has a opening and closing element so that the properties of the Bean can be set. You'll notice that the Bean only has one property set, but there could be several.

Now that the values are set, it is just a matter of using jsp:getProperty and the browser displays the value.

```
<HTML>
  <HEAD>
    <TITLE>Bean Example</TITLE>
  </HEAD>
<BODY>
```

```
<H2>XPlatform Bean Example</H2>
<jsp:useBean id="getQuote"
             class="XPlatform.getServiceBean">
<jsp:setProperty name="getQuote"
                 property="symbol"
                 value="C" />
</jsp:useBean>

 Value:<jsp:getProperty name="getQuote"
                        property="price"/>
</BODY>
</HTML>
```

The Bean code still has much of the same Web Service code in it that has been used throughout the examples found in this chapter, but now it appears within a get method named getPrice. By using standard get and set methods the JSP easily communicates with the Bean. Notice that there aren't any getSymbol or setPrice methods because they are not needed for this example. This is the complete code example for the Bean.

```
//bean must appear in WebInf/classes/XPlatform
package XPlatform;
import java.net.*;
import java.util.*;
import org.apache.soap.*;
import org.apache.soap.rpc.*;

public class getServiceBean {

  public String symbol = null;
  public String price = null;

  public void setSymbol(String s) {
    symbol = s;
  }

  public String getPrice() {

    String encodingStyleURI =
    Constants.NS_URI_SOAP_ENC;
    URL url = null;
    //set up URL to call
```

```
      try {
        url = new URL
       ("http://homer:8080/soap/servlet/rpcrouter");
      } catch (MalformedURLException e){
       System.out.println("Error:" + e);
     }

      //set up to the call to the Web Service
      Call call = new Call ();
      call.setTargetObjectURI ("urn:simple-stock-quote");
      call.setMethodName ("getTestQuote");
      call.setEncodingStyleURI(encodingStyleURI);
      Vector params = new Vector ();
      params.addElement (new Parameter("symbol",
      String.class, symbol, null));
      call.setParams (params);
      Response resp;
      String returnValue = null;


      //try to call the Web Service, if successful return
      //the value
      try {
         resp = call.invoke ( url, "" );
         if (resp.generatedFault ()) {
            Fault fault = resp.getFault ();
            System.out.println("fault: " + fault );
            price = "-1";
         } else {
            Parameter result = resp.getReturnValue ();
            price = result.getValue().toString();
         }


      } catch (SOAPException e) {
         System.out.println("Error:" + e);
      }
      return(price);
   }
}
```

Figure 7.11 shows the output of the JSP/Bean example

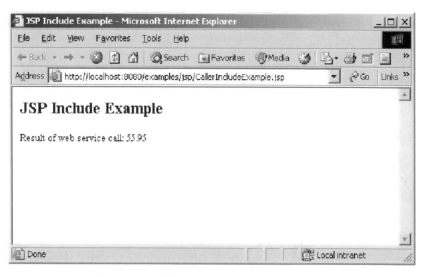

FIGURE 7.11   The output of the JSP/Bean example.

CONCLUSION

The Apache *SOAP* library offers some of the functionality found in Microsoft's *.NET* implementation. There are classes and methods available to create and consume Web Services. There is a server to use and deploy Web Services to, and the code allows you to use a variety of consumers.

The main difference is that we don't have the Web interface like we do for a service *.NET*. In addition, WSDL isn't generated automatically and there aren't the convenient features for explaining what each service and method does.

Chapter 9 covers Apache *Axis*, which is the next generation of Web Services from the Apache Group. This is a complete rewrite of the Apache *SOAP* library, and you'll find that many of the shortcomings found in this chapter disappear when using *Axis*.

8 Web Services with Apache *Axis*

In This Chapter

- Prerequisites for Using Apache *Axis*
- Creating and Deploying an Axis Web Service: Differences between Apache *Axis* and *SOAP*
- Creating a Consumer with *Axis*

Apache *Axis*™ is an attempt by the Apache Group to create an open source Web Services tool that is compatible with other available systems such as Microsoft's *.NET*. Several features that were missing from Apache *SOAP* appear in this release.

This release is a complete rewrite of Java Web Services. Apache *Axis* has its roots in IBM's Web Services toolkit. At one point, the project transferred from IBM to the Apache Group.

It is important to note that this chapter is based on a Beta release from the Summer of 2002. It is possible that there will be differences between the examples generated with the Beta version in this chapter that may end up having some incompatibilities with the release version that comes out in the Fall of 2002.

Some of the best features found in *Axis* deal with creating WSDL. Apache *Axis* makes WSDL easily available; you'll see in Chapter 9 that this makes it quite easy to interact with *.NET*. The WSDL tools also generate Java proxy code that makes it much easier to write a client. These WSDL tools are similar to Microsoft's WSDL tool discussed in Chapter 6.

This chapter begins by discussing the prerequisites for using Apache *Axis* on your system.

PREREQUISITES FOR USING APACHE *AXIS*

Unlike software you purchase, open source software usually requires several installation steps before you are able to utilize the software. Apache *Axis* is no different, but you'll find the setup slightly easier than Apache *SOAP*.

Apache *Axis* utilizes *Tomcat* and Chapter 7 discusses how to install it. The only difference with *Axis* is that you do not need to change the CLASSPATH settings of *Tomcat*'s startup script.

The other steps for the installation of *Axis* include installing the appropriate libraries, setting up the CLASSPATH, and then testing.

Installing *Axis* and *Xerces*

Appendix B provides the URL for downloading *Axis* from the Apache Group. Once you have the zip file containing the *Axis* distribution, extract it into C:\xmlapache just like what you did in Chapter 7. Go into the `axis-1_0\webapps` directory under `c:\xmlapache` and take the *Axis* directory and move it into *Tomcat*'s `webapps` directory. This will install the *Axis* webapp under "*Tomcat.*"

For *Axis* to handle XML, you need to put *xerces.jar* (the Apache group's XML parser) in the *Axis* webapps `lib` directory. For example, if you installed everything under the recommended path `c:\xmlapache`, the entire path for *xerces.jar* would be this: `c:\xmlapache\tomcat\webapps\axis\lib\xerces.jar`. Note that for the examples in this chapter, *Xerces* Version 1.4.4 was used—which is the same version that Apache *SOAP* requires.

Once the *Axis* webapps directory is installed and the *xerces.jar* is placed in the appropriate directory, the next step involves setting up the CLASSPATH and other environmental variables.

Setting up the Environment

For Apache *Axis* to execute, you need to modify the environment so that it can find the Java executables and libraries it needs.

The first environmental variables needed are for the location of the executables needed to compile and run Java programs and the variable needed to tell *Tomcat* where the Java directory resides.

The following example shows a MS DOS batch file that sets up the environment. The first line set adds the location of the Java executables to the path. The second example line sets the JAVA_HOME variable for *Tomcat*. This simply needs the path to the Java *SDK*. The CLASSPATH variable tells Java where to find certain critical libraries. Each of the jar files found in `c:\xmlapache\tomcat\webapps\axis\WEB-INF\lib\` needs to be added to the CLASSPATH in order for *Axis* to function properly.

```
set PATH=%PATH%;c:\jdk1.4.0\bin;.
set JAVA_HOME=c:\jdk1.4.0
set CLASSPATH=
  c:\xmlapache\tomcat\webapps\axis\WEB-INF\lib\axis.jar;
  c:\xmlapache\tomcat\webapps\axis\WEB-INF\lib\commons-
  logging.jar;
  c:\xmlapache\tomcat\webapps\axis\WEBINF\lib\jaxrpc.jar;
  c:\xmlapache\tomcat\webapps\axis\WEB-INF\lib\saaj.jar;
  c:\xmlapache\tomcat\webapps\axis\WEB-INF\lib\log4j-1.2.4.jar;
  c:\xmlapache\tomcat\webapps\axis\WEB-INF\lib\tt-bytecode.jar;
  c:\xmlapache\tomcat\webapps\axis\WEB-INF\lib\xerces.jar;
```

```
        c:\xmlapache\tomcat\webapps\axis\WEB-INF\lib\wsdl4j.jar;
c:\xmlapache\axis-1_0\;
```

Executing this bat file and then starting *Tomcat* will allow the *Axis* examples to function properly. Note that this bat file can be found on the CD-ROM in the directory for this chapter.

Testing the Installation of Apache *Axis*

Once *Tomcat* is started, you should see the *Axis* directory load with all the other webapp directories. Figure 8.1 shows the Apache *Axis* home page as it appears in *Internet Explorer*.

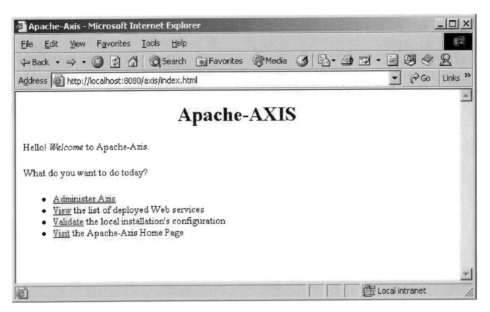

FIGURE 8.1 The Apache *Axis* home page shown in *Internet Explorer*.

One feature that comes with Apache *Axis* that didn't appear in the *SOAP* library is the diagnostic tool, and this is found from the "Validate" link on the home page shown in Figure 8.1. This tells you if *Axis* is finding all the libraries it needs in order to operate properly. Figure 8.2 shows the results of the diagnostics.

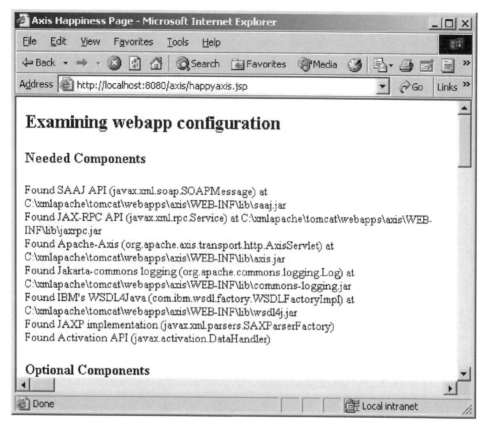

FIGURE 8.2   The results of the Apache *Axis* diagnostics.

The diagnostics should report that there are two missing optional libraries. You don't need to worry about these yet because the examples in this chapter don't use the missing libraries.

Axis can also list the available services on a particular system. Just follow the "View Deployed Web Services" link on the *Axis* home page. Figure 8.3 shows the available services on a sample installation of *Axis*.

Another quick check of the *Axis* installation is to execute one of the examples that come with the *Axis* distribution. Open a DOS prompt and change into the c:\xmlapache\axis-1_0\stock directory. Now execute testit.cmd. This executes a DOS batch program that tests *Axis*, installs a real stock service, tests the stock

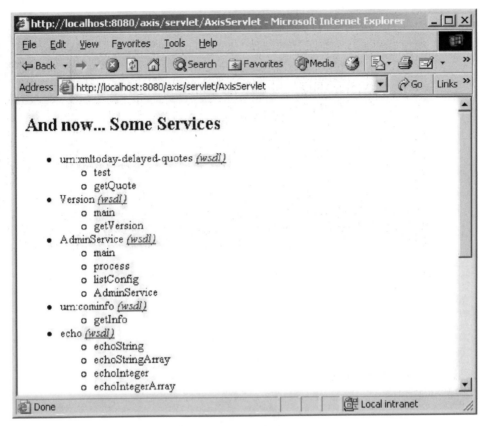

FIGURE 8.3 How Apache *Axis* lists installed Web Services.

service, and then uninstalls the service. Figure 8.4 shows the output that `testit.cmd` generates.

If a lot of Java errors appear in the DOS window when executing `testit.cmd`, *the most likely culprit is that your* CLASSPATH *does not have all the correct libraries.*

Now that you have installed and tested Apache *Axis*, the next step is to actually create your own Web Service.

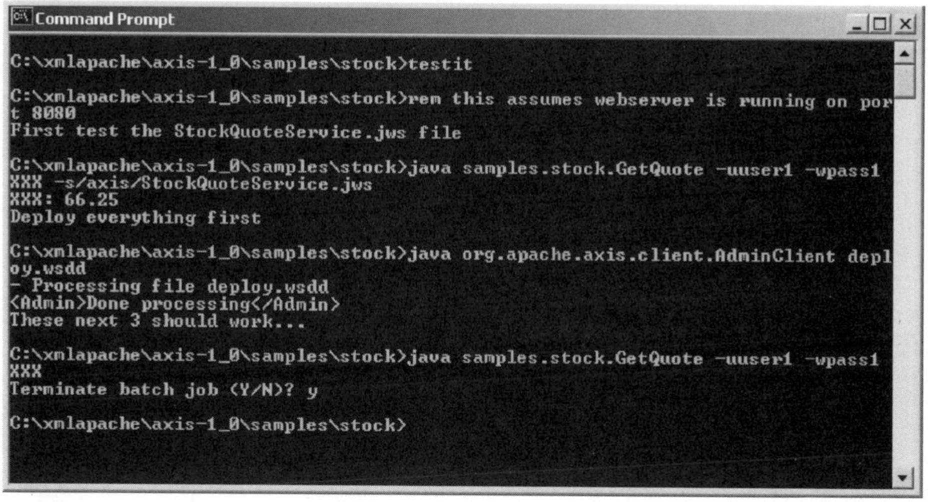

FIGURE 8.4 The results of running the `testit.cmd` file in the Apache Group's stock example.

CREATING AND DEPLOYING AN *AXIS* WEB SERVICE: DIFFERENCES BETWEEN APACHE *AXIS* AND *SOAP*

The coding of a Web Service with *Axis* doesn't look that much different than a *SOAP* Web Service shown in Chapter 7. The main differences are the import statements and how you compile the service for deployment. The following examples show an *Axis* Web Service and its XML deployment descriptor.

The Web Service

Consider the following code for *Axis* that contains the code for the `SimpleStockExample` class used in previous chapters.

```
public class SimpleStockExample {

    public float getTestQuote(String symbol)
        throws Exception {
```

```
if ( symbol.equals("C") ) {
   return( (float) 55.95 );
} else {
   return( (float) -1);
}

}

}
```

This is just a simple program with no import or package statements. You can compile this in the directory where you created it to ensure that it is syntactically correct, but when you're ready to deploy, change the extension of the Java (Simple-StockExample.java) file from .java to .jws (SimpleStockExample.jws) and copy it into c:\xmlapache\tomcat\webapps\axis\. With *Tomcat* running, point your browser to the following URL: http://localhost:8080/axis/SimpleStockExample. By pointing the browser to that URL, you compile the code into a Web Service and the results look like Figure 8.5.

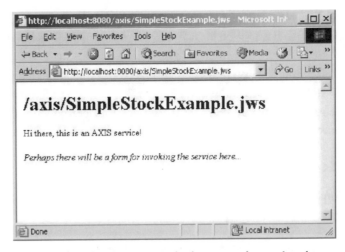

FIGURE 8.5 The response to the browser when pointed to the SimpleStockExample Web Service.

Apache Axis *creates its own WSDL for each Web Service, and just like Microsoft's* .NET, *this WSDL is available by simply putting "WSDL" in the query string of the URL like this:* http://localhost:8080/axis/SimpleStockExample?WSDL.

If, for some reason, this method of deployment doesn't work for you, you can create a directory under the *Axis* directory in the *Tomcat* directory structure and create a Web Service that looks like the following.

```
package samples.SimpleStock;

    import org.w3c.dom.Document;
    import org.w3c.dom.Element;
    import org.w3c.dom.NodeList;

    import javax.xml.parsers.DocumentBuilder;
    import javax.xml.parsers.DocumentBuilderFactory;

    public class SimpleStockExample {

    public float getTestQuote(String symbol)
        throws Exception {

      if ( symbol.equals("C") ) {
         return( (float) 55.95 );
      } else {
         return( (float) -1);
      }

    }

    }
```

This last example is more like what was found with the Apache *SOAP* library, and also requires a deployment descriptor. Using this method may be more work, but it may give you more flexibility because a deployment descriptor allows you to specify certain meta information that you don't get when using a .jws file.

The Deployment Descriptor

The deployment descriptor in Apache *Axis* uses different syntax than its predecessor in *SOAP*, but the idea remains the same. It's an XML document that describes how the *Axis* server should deploy the service if it is necessary to specify additional information beyond just creating a .jws file.

In the following example, note that the deployment element starts off by giving this deployment description a name and describes the namespaces used in the document. The service element attributes name the service and describe the type or provider that makes the service available. The child elements, which are both parameter, describe the class available to the service and the methods to be accessed. In this case "*" gets used to indicate that all the methods are available.

```
<deployment
    name="load"
    xmlns="http://xml.apache.org/axis/wsdd/"
    xmlns:java=
    "http://xml.apache.org/axis/wsdd/providers/java">

  <service name="SimpleStockExample"
           provider="java:RPC">
    <parameter name="className"
               value=
               "samples.SimpleStock.SimpleStockExample"/>
    <parameter name="allowedMethods" value="*"/>
  </service>
</deployment>
```

Save the XML in a file called deploy.wsdd and use the following command to set up the Web Service with *Axis*:

```
java org.apache.axis.client.AdminClient deploy.wsdd
```

Now, if you go look at the deployed Web Services in *Axis'* admin GUI, you should see that SimpleStockExample is now considered a deployed Web Service, as shown in Figure 8.6.

In Apache *SOAP* there was no easy means of providing WSDL, but with *Axis* the WSDL for the example is found by adding "WSDL" in the query string like this:

```
http://localhost:8080/axis/services/SimpleStockExample?wsdl
```

Figure 8.7 shows the WSDL output for the SimpleStockExample class.

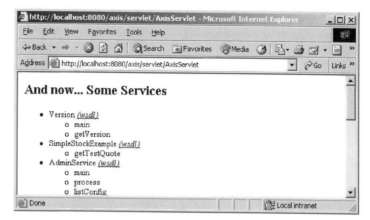

FIGURE 8.6   All the deployed Web Services available at
http://localhost:8080/axis/AxisServlet.

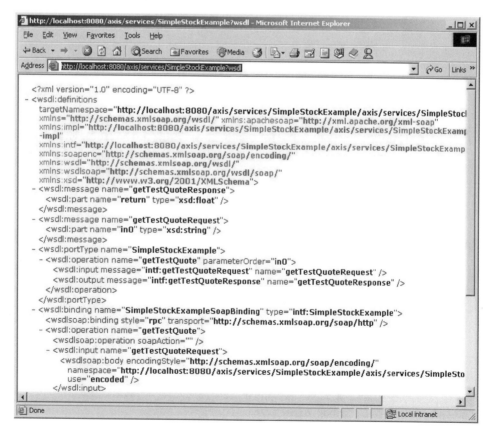

FIGURE 8.7   The WSDL for the `SimpleStockExample` from Apache *Axis*.

If you look closely at the WSDL for the `SimpleStockExample`, *you notice that many of the namespace entries are actually incorrect. Entries for this service should use the URL:* http://localhost:8080/axis/services/SimpleStockExample *but are, instead, using* http://localhost:8080/axis/services/SimpleStockExample/axis/services/SimpleStockExample. *Remember that Axis was in Beta when this chapter was written so this bug might get fixed by the final release.*

Undeploying the `SimpleStockExample`

An XML file is also required to remove or undeploy a Web Service from the *Axis* server.

The following XML example undeploys the `SimpleStockExample`. The root element is `undeployment`, whose attributes define the name of this undeployment descriptor and the namespace for the XML file. The `service` child element describes which class should no longer be served by *Axis*.

```
<undeployment name="load"
              xmlns="http://xml.apache.org/axis/wsdd/">
    <service name="SimpleStockExample"/>
</undeployment>
```

Then, by executing the following command, the `AdminClient` reads the `undeploy.wsdd` to determine which class or classes to remove.

```
java org.apache.axis.client.AdminClient undeploy.wsdd
```

Relaxing Deployment Rules

Unlike Apache *SOAP*, *Axis* does not come with remote administration installed. It only allows you to administer from the local machine. If you wish to allow remote administration to occur so an entire development team has access to deployment, the `server-confi.wsdd` file needs to be modified. This file contains directives for the entire *Axis* server, and by adding the following to the `AdminService` directive, the deployment of Web Services from remote machines is enabled.

```
<service name="AdminService" provider="java:MSG">
    ...Other directives
```

```
    <parameter name="enableRemoteAdmin" value="true"/>
</service>
```

The entire entry might look like the following.

```
<service name="AdminService" provider="java:MSG">
  <parameter name="allowedMethods" value="AdminService"/>
  <parameter name="enableRemoteAdmin" value="false"/>
  <parameter name="className" value="org.apache.axis.utils.Admin"/>
  <parameter name="sendXsiTypes" value="true"/>
  <parameter name="sendMultiRefs" value="true"/>
  <parameter name="sendXMLDeclaration" value="true"/>
  <namespace>http://xml.apache.org/axis/wsdd/</namespace>
  <parameter name="enableRemoteAdmin" value="true"/>
</service>
```

Having remote administration rights turned on by default was a major security problem with the *SOAP* library because many people wouldn't know it was turned on. This way users cannot deploy Web Services to your server without your knowledge.

Using *TCP Monitor*

TCP Monitor™ is an updated version of the *TCP Tunnel* GUI discussed in Chapter 7. It is a tool that allows you to view the requests and responses to a particular server. It acts as a proxy so that it intercepts requests and responses, prints them to the screen, and then forwards them to the appropriate server or client. To use this tool, execute the following command.

```
java org.apache.axis.utils.tcpmon 8000 localhost 8080
```

Figure 8.8 displays the *TCP Monitor* when it first loads under *Windows 2000*. Click on the tab that says "Port 8080" and you'll see any Web Service request and responses to that port. Figure 8.9 shows *TCP Montior* watching Port 8080.

Figure 8.10 shows a request and response within the *TCP Monitor*.

Unlike the *TCP Tunnel* GUI, *TCP Monitor* handles requests for multiple ports and, therefore, you can monitor several different Web Services.

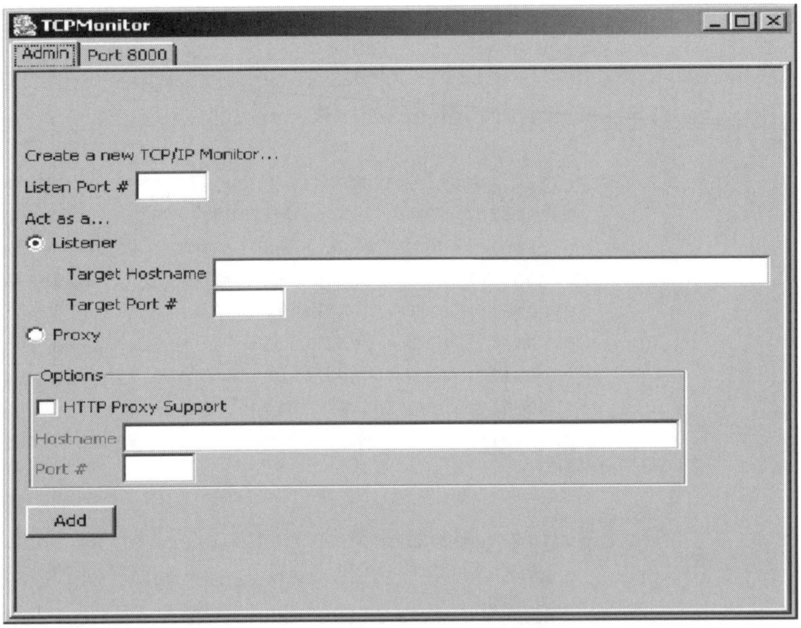

FIGURE 8.8 The *TCP Monitor* as it appears when it first loads.

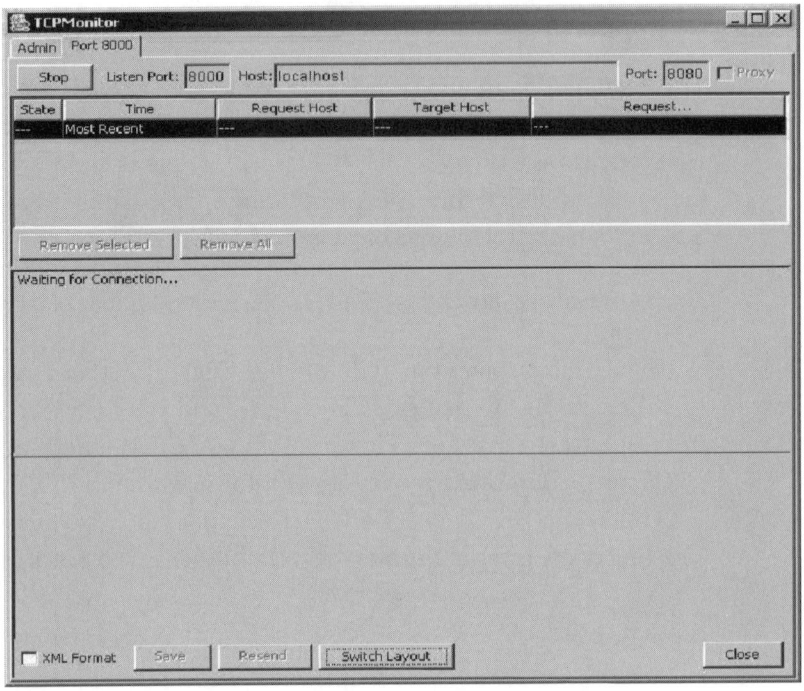

FIGURE 8.9 The "Port 8080" display window of *TCP Monitor*.

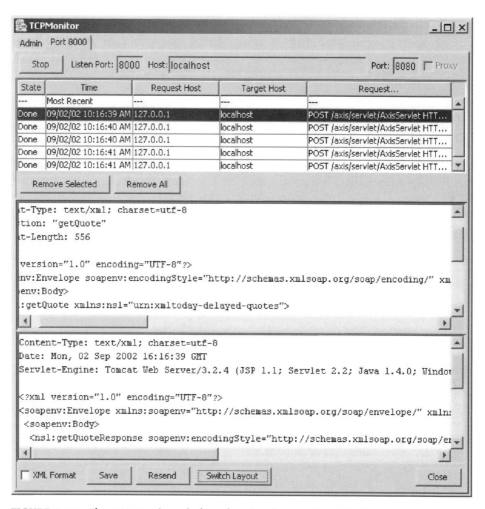

FIGURE 8.10   The *TCP Monitor* window showing the results of *SOAP* requests and responses.

CREATING A CONSUMER WITH *AXIS*

Creating consumers with *Axis* is really not that much different than the examples in Chapter 7. There are some different include files and a tool to create proxy code much like the WSDL tool found in *.NET*, but the concept of creating consumers is the same.

In this chapter, we focus on creating a simple command line consumer and the same program that creates a proxy. Using Axis within JSP and other consumers is covered in Chapter 9. which is a case study of using *.NET* and Java Web Services together.

Command Line Application

The following code is a simple command line application that invokes the *Axis* Web Service example shown earlier in the chapter. The example starts off by defining all the different Java libraries needed for this example with the import statements. Then there is a class definition followed by the definition of the main method.

Then the example begins to call the Web Service by reading the options passed in from the command line, creating a call to the URL where the Web Service resides, defining the class and method to call from the server, defining the result, and finally printing the result.

```java
import org.apache.axis.AxisFault;
import org.apache.axis.client.Call;
import org.apache.axis.client.Service;
import org.apache.axis.encoding.XMLType;
import org.apache.axis.utils.Options;

import javax.xml.rpc.ParameterMode;

import javax.xml.namespace.QName;
import java.net.URL;

public class getSimpleStock {
    public static void main (String[] args)
    throws Exception {

  String symbol = null;
  //get command line args
  Options  myOpts     = new Options( args );

  args = myOpts.getRemainingArgs();

  //begin a call to a Web Service
  Service  myService = new Service();
  Call     myCall     = (Call) myService.createCall();
```

```
          //location of axis server
          myOpts.setDefaultURL(
          "http://localhost:8000/axis/servlet/AxisServlet" );


          myCall.setTargetEndpointAddress
          ( new URL(myOpts.getURL()) );

          myCall.setUseSOAPAction( true );
          //the method to call
          myCall.setSOAPActionURI( "getTestQuote" );
          //how to encode the request/response
          myCall.setEncodingStyle
          ("http://schemas.xmlsoap.org/soap/encoding/" );
          //define the class and method used
          myCall.setOperationName( new
          QName("SimpleStockExample", "getTestQuote") );
          //add symbol to the request
          myCall.addParameter
          ( "symbol", XMLType.XSD_STRING, ParameterMode.IN );
          myCall.setReturnType( XMLType.XSD_FLOAT );
          //make the actual object call
          Object myResult =
          myCall.invoke( new Object[] { symbol = args[0] } );
          //print the result
          System.out.println("This is the returned value: " +
          ((Float)myResult).floatValue());
        }
    }
```

If you compare this to the examples in Chapter 6, this is a lot of code to write to call a Web Service when compared to *.NET*—especially since this code needs to be used each time one particular method gets called.

To make calling a Web Service easier with Apache tools, *Axis* comes with *WSDL2Java* application that generates Java proxy code to make calling the Web Service easier and quicker. The next section covers the creation of this proxy.

Using *WSDL2Java* to Create a Proxy

With *Axis*, WSDL is generated automatically and can be seen by simply putting "wsdl" in the query string. Consider the following URL: *http://localhost:8080/axis/ services/SimpleStockExample?wsdl*

This will display the WSDL for the `SimpleStockExample` Web Service shown earlier in this chapter. Figure 8.11 shows how the WSDL appears in *Internet Explorer*.

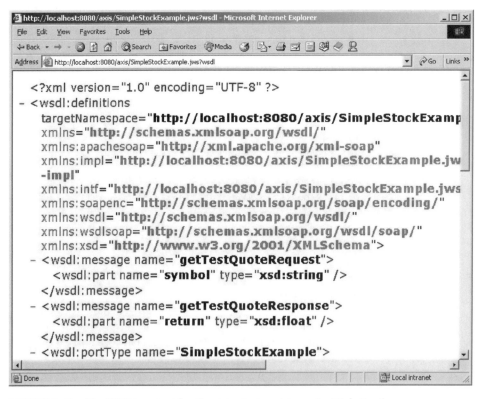

FIGURE 8.11 The WSDL output for the `SimpleStockExample` Web Service.

To create a Java proxy for a client to use, execute *WSDL2Java* and send the URL of the `SimpleStockQuote`'s WSDL output. The command should look something like the following.

```
java org.apache.axis.wsdl.WSDL2Java
http://localhost:8080/axis/services/SimpleStockExample?wsdl
```

This creates a directory called `localhost`. Within that directory you will find the following four files.

```
SimpleStockExample.javaSimpleStockExampleSoapBindingStub.java
SimpleStockExampleService.java  SimpleStockExampleServiceLocator.java.
```

`SimpleStockExample` and `SimpleStockExampleService` are both interfaces that define the functionality that the other two Java files must implement. `SimpleStock-ExampleServiceLocator` implements SimpleStockExampleService and defines the location of the Web Service and methods.

`SimpleStockExampleSoapBindingStub.java` implements `SimpleStockExample-ServiceLocator.java` and handles the code related to the call.

The following is the code for `SimpleStockExample.java`.

```
/**
 * SimpleStockExample.java
 *
 * This file was auto-generated from WSDL
 * by the Apache Axis WSDL2Java emitter.
 */

package localhost;

public interface SimpleStockExample extends
    java.rmi.Remote {
    public float getTestQuote(java.lang.String in0)
    throws java.rmi.RemoteException;
}
```

Notice how the code simply defines an interface, and notice how `SimpleStock-ExampleServiceLocator.java` utilizes this interface.

```
/**
 * SimpleStockExampleServiceLocator.java
 *
 * This file was auto-generated from WSDL
 * by the Apache Axis WSDL2Java emitter.
 */

package localhost;

public class SimpleStockExampleServiceLocator extends
org.apache.axis.client.Service implements
 localhost.SimpleStockExampleService {
```

```java
// Use to get a proxy class for SimpleStockExample
private final java.lang.String
SimpleStockExample_address =
"http://localhost:8080/axis/services/SimpleStockExample";

public String getSimpleStockExampleAddress() {
    return SimpleStockExample_address;
}

public localhost.SimpleStockExample
getSimpleStockExample() throws
javax.xml.rpc.ServiceException {
   java.net.URL endpoint;
   try {
       endpoint = new java.net.URL
       (SimpleStockExample_address);
   }
   catch (java.net.MalformedURLException e) {
       return null;
   }
   return getSimpleStockExample(endpoint);
}

public localhost.SimpleStockExample
getSimpleStockExample(java.net.URL portAddress) throws
javax.xml.rpc.ServiceException {
    try {
        return new
        localhost.SimpleStockExampleSoapBindingStub
        (portAddress, this);
    }
    catch (org.apache.axis.AxisFault e) {
        return null; // ???
    }
}

/**
 * For the given interface, get the stub
 * implementation.
 * If this service has no port for the given interface,
 * then ServiceException is thrown.
 */
public java.rmi.Remote getPort(Class
   serviceEndpointInterface) throws
```

```
 javax.xml.rpc.ServiceException {
   try {
 if
 (localhost.SimpleStockExample.class.isAssignableFrom
 (serviceEndpointInterface)) {
            return new
  localhost.SimpleStockExampleSoapBindingStub
 (new java.net.URL(SimpleStockExample_address), this);
        }
    }
    catch (Throwable t) {
        throw new javax.xml.rpc.ServiceException(t);
    }
    throw new javax.xml.rpc.ServiceException("There is
    no stub implementation for the interface:  " +
    (serviceEndpointInterface == null ? "null" :
    serviceEndpointInterface.getName()));
 }

 }
```

The previous Java code is mainly centered on finding the name of the service. The following code snippet is `SimpleStockExampleService.java`. This defines the interface for the actually getting values from the Web Service.

```
/**
 * SimpleStockExampleService.java
 *
 * This file was auto-generated from WSDL
 * by the Apache Axis WSDL2Java emitter.
 */

package localhost;

public interface SimpleStockExampleService extends
javax.xml.rpc.Service {
public String getSimpleStockExampleAddress();

public localhost.SimpleStockExample
getSimpleStockExample() throws
javax.xml.rpc.ServiceException;

public localhost.SimpleStockExample
```

```
getSimpleStockExample(java.net.URL portAddress) throws
javax.xml.rpc.ServiceException;
}
```

Finally, `SimpleStockExampleSoapBindingStub.java` contains the code defined in the previous interface. This code focuses on creating the call and getting values back from the Web Service.

```java
/**
 * SimpleStockExampleSoapBindingStub.java
 *
 * This file was auto-generated from WSDL
 * by the Apache Axis WSDL2Java emitter.
 */

package localhost;

public class SimpleStockExampleSoapBindingStub extends
org.apache.axis.client.Stub implements
localhost.SimpleStockExample {
private java.util.Vector cachedSerClasses
= new java.util.Vector();
private java.util.Vector cachedSerQNames
= new java.util.Vector();
private java.util.Vector cachedSerFactories
= new java.util.Vector();
private java.util.Vector cachedDeserFactories
= new java.util.Vector();

public SimpleStockExampleSoapBindingStub() throws
org.apache.axis.AxisFault {
    this(null);
}

public SimpleStockExampleSoapBindingStub
(java.net.URL endpointURL,
javax.xml.rpc.Service service) throws
org.apache.axis.AxisFault {
    this(service);
    super.cachedEndpoint = endpointURL;
}

public SimpleStockExampleSoapBindingStub
```

```
(javax.xml.rpc.Service service) throws
 org.apache.axis.AxisFault {
     try {
         if (service == null) {
             super.service
             = new org.apache.axis.client.Service();
         } else {
             super.service = service;
         }
     }
     catch(java.lang.Exception t) {
         throw org.apache.axis.AxisFault.makeFault(t);
     }
 }

 private org.apache.axis.client.Call createCall() throws
 java.rmi.RemoteException {
     try {
         org.apache.axis.client.Call call =
         (org.apache.axis.client.Call)
         super.service.createCall();
         if (super.maintainSessionSet) {

    call.setMaintainSession(super.maintainSession);
         }
         if (super.cachedUsername != null) {
             call.setUsername(super.cachedUsername);
         }
         if (super.cachedPassword != null) {
             call.setPassword(super.cachedPassword);
         }
         if (super.cachedEndpoint != null) {

       call.setTargetEndpointAddress
      (super.cachedEndpoint);
         }
         if (super.cachedTimeout != null) {
             call.setTimeout(super.cachedTimeout);
         }
         java.util.Enumeration keys =
         super.cachedProperties.keys();
         while (keys.hasMoreElements()) {
             String key = (String) keys.nextElement();
```

```java
                    if(call.isPropertySupported(key))
                        call.setProperty(key,
                        super.cachedProperties.get(key));
                    else
                        call.setScopedProperty(key,
                        super.cachedProperties.get(key));
                }
                // All the type mapping information is
                //registered
                // when the first call is made.
                // The type mapping information is actually
                //registered in
                // the TypeMappingRegistry of the service,
                //which
                // is the reason why registration is only
                //needed for the first call.
                synchronized (this) {
                    if (firstCall()) {
                        // must set encoding style before
                        //registering serializers

                      call.setEncodingStyle
                      (org.apache.axis.Constants.URI_SOAP11_ENC);
                            for (int i = 0; i <
                                 cachedSerFactories.size(); ++i) {
                                Class cls = (Class)
                                cachedSerClasses.get(i);
                                javax.xml.namespace.QName qName =
                                (javax.xml.namespace.QName)
                                cachedSerQNames.get(i);
                                Class sf = (Class)
                                cachedSerFactories.get(i);
                                Class df = (Class)
                                cachedDeserFactories.get(i);
                                call.registerTypeMapping
                                (cls, qName, sf, df, false);
                            }
                        }
                    }
                    return call;
            }
            catch (Throwable t) {
                throw new org.apache.axis.AxisFault
```

```
            ("Failure trying to get the Call object", t);
        }
    }
more and more lines of code ...
}
```

The last example code was cut short for a very good reason. You don't really need to know what the code is doing, much like you don't need to worry about the proxy code in *.NET*.

Once the Java files are created, compile them with the following command within the `localhost` directory:

```
javac *.java
```

This should make all the corresponding class files and compile without error. Now the classes are available to any client you wish to create. The following example utilizes these classes.

The import statement brings in all the classes that *WSDL2Java* created and you compiled. Then the class `getSimpleStockWSDL` is created with a main method. The result variable is defined so it handles a value returned by the method. Next, the code creates the object `myService`, which contains information about where the object and method you need to call reside. The next step creates the object `mySOAP`, which is the actual object that represents the `SimpleStockQuote` class. Notice that the next step is the `mySOAP` object calling the `getTestQuote` method. Then the value of `result` is output. This piece of code is the simplest call to a Java Web Service shown in this book so far.

```
import localhost.*;

public class getSimpleStockWSDL {
    public static void main(String [] args) throws
    Exception {
        //The type the service returns.
        double result;
        SimpleStockExampleServiceLocator myService =
        new  SimpleStockExampleServiceLocator();

        localhost.SimpleStockExample mySOAP =
        myService.getSimpleStockExample();
```

```
        result = mySOAP.getTestQuote("C");

    System.out.println
    ("This is the value: " + result);
    }
}
```

Doesn't this seem a little like using the WSDL tool found in .NET? It should. Microsoft and IBM collaborated on the WSDL standard, and Apache Axis has its roots as a project that began at IBM. So it's no wonder that the two processes are similar.

CONCLUSION

When Web Services and *SOAP* were emerging technologies, Microsoft really took the lead in making the technology a reality. For Java to have access to Web Services, many third-party vendors rushed Java Web Services technology to the market. For a while, using a third-party tool was the only option for a Java developer unless they wanted to create their own implementation.

With Apache *Axis*, many more opportunities are opened for the Java developer. Instead of being forced to purchase a tool, one can simply download Apache *Axis* and see if Web Services fit into the infrastructure or solve their problems. The fact that *Axis* provides tools to generate WSDL makes it easy for Java-based Web Services to communicate with Microsoft's *.NET* initiative. How to perform this integration is covered in Chapter 9.

Web Service Applications

Now that you understand the basics and have seen some Web Service implementations, it's time to examine some of the more practical problems involved with Web Services.

Most large corporations operate in a mixed environment where both *Windows* machines and some flavor of UNIX coexist with one another. One of the more difficult problems is getting these two platforms to communicate without having to write your own software. Web Services can act as a bridge between these two technologies.

Section III examines how to get different Web Service implementations to communicate, discusses how to secure these transactions, studies a more complex example of disparate systems communicating, and examines what it would take for you to create your own Web Service implementation.

9

Java and *.NET* Web Service Integration

In This Chapter

- Three Test Web Services
- Testing Web Services
- Cross-Platform Consumers
- Third-Party Tools and Other Languages

Having objects that communicate between diverse systems, such as *.NET* and Java, gives a developer a great deal of flexibility when developing remote objects. Someone creating a Business to Business (B2B) infrastructure can choose the technology that best fits the existing infrastructure without worrying about excluding a potential customer because of technology choice. Even though *.NET* and the Apache Web Service products communicate together relatively easily, you still want to engage in a great deal of testing with a customer looking for potential problems or points of failure. For instance, security under Apache *Axis* works differently under *.NET*, so you can't jump into Web Services without knowing who your audience is and ensuring that you test your code with all appropriate consumers.

Another point to consider in this chapter is how you access Web Services. In the examples in this book, all the Web Services reside on the same machine that includes code in both Java and C#. This would not happen in a real installation because C# code would execute on a *Windows* server whereas Java code would be executing under Linux or UNIX. Although it is possible to configure IIS to proxy requests to *Tomcat*, most server-side Java implementations occur on a server with some form of UNIX. Figure 9.1 illustrates how you execute Web Services and consumers for the examples in this chapter.

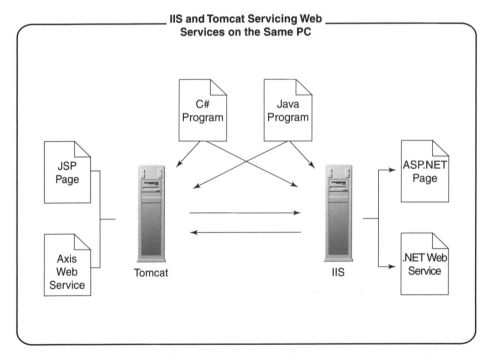

FIGURE 9.1   Examples in this book execute under different Web Servers but still reside on the same machine and OS.

Figure 9.2 shows how a deployment might look like in an Enterprise Setting.

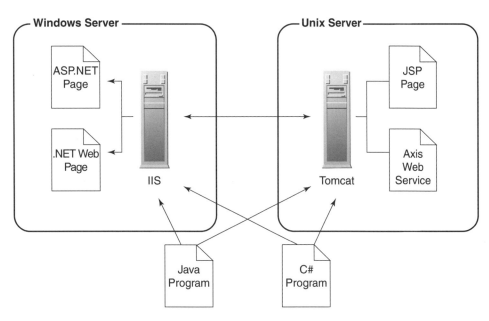

FIGURE 9.2   Web Services for C# and the Apache group residing on different machines where UNIX is the most likely platform for Apache products.

In this chapter, there are examples from all three platforms covered so far: *.NET*, Apache *SOAP*, and Apache *Axis*. Apache *SOAP* was the first implementation of Java Web Services by the Apache Group, and thus being in the marketplace longer than *.NET* there is a chance you'll come across a legacy implementation. That's why this chapter includes examples of integrating Apache *SOAP* with the other technologies. The chapter begins by introducing three Web Services that utilize the different frameworks.

THREE TEST WEB SERVICES

This section shows three different Web Services that identify the Web Service implementation called. This shows that consumers of various types are able to consume Web Services from Apache and Microsoft. Once the three Web Services are created, examples later in the chapter, written in both C# and Java, call all the different types of services.

.NET Web Service

The following Web Service was created in *Visual Studio.NET* and simply returns the text "This is Microsoft DOT Net." Even though this Web Service is written in C#, you'll find that the Java consumers shown later in the chapter are still able to call it with relative ease.

The URL of this Web Service on the author's machine is
http://localhost/XPlatform/MSNETID/Service1.asmx?op=ServiceId

Just like the other *.NET* Web Service examples shown in Chapter 6, *Visual Studio.NET* generates much of the code. Notice that the WebService and WebMethod tags are added for further clarification to the consumer. The only other modification needed is to add the method that returns the identifying text.

```csharp
using System;
using System.Collections;
using System.ComponentModel;
using System.Data;
using System.Diagnostics;
using System.Web;
using System.Web.Services;

namespace MSNETID
{
[WebService(Description="Web Service that informs a consumer that
           this is .NET",
           Namespace="http://www.advocatemedia.com/")]

  public class Service1 : System.Web.Services.WebService
  {
      public Service1()
      {
          //CODEGEN: This call is required by the ASP.NET Web
             //Services Designer
          InitializeComponent();
      }

      #region Component Designer generated code
```

```
//Required by the Web Services Designer
private IContainer components = null;

private void InitializeComponent()
{
}


/// <summary>
/// Clean up any resources being used.
/// </summary>
protected override void Dispose( bool disposing )
{
    if(disposing && components != null)
    {
        components.Dispose();
    }
    base.Dispose(disposing);
}

#endregion

[WebMethod
  (Description="This method just returns that this is MS
  .NET")]
public string ServiceId()
{
    return "This is Microsoft DOT NET";
}
   }
}
```

Figure 9.3 displays the output of this Web Service in *Internet Explorer*.

The following XML, which is obtained by testing the Web Service in a browser, demonstrates that this Web Service is a Microsoft technology.

```
<?xml version="1.0" encoding="utf-8" ?>
  <string xmlns="http://www.advocatemedia.com/">
      This is Microsoft DOT NET
  </string>
```

Now that you have a Microsoft Web Service, you need Apache *SOAP* and *Axis* Web Services in order to demonstrate cross-platform compatibility.

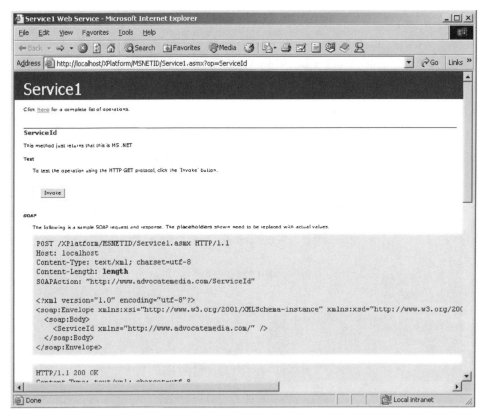

FIGURE 9.3 The MSNETID Web Service output to *Internet Explorer*.

Apache *SOAP* Web Service

Even though Apache *SOAP* is obsolete with the advent of *Axis*, it is still important to study how to consume a *SOAP* Web Service in case you come across a legacy installation.

This example is similar to the one found in Chapter 7, but instead of a simple stock quote the text "This is Apache SOAP" returns to the client. The service uses the standard import statements and is part of the package samples.javaid. The file needs to be named ApacheSoapId because this is the name of the class defined in the file. The only method defined is ServiceId, and this returns text to identify that this is a *SOAP* service.

```
package samples.javaid;

import java.net.URL;
import java.io.*;
import org.w3c.dom.*;
import org.xml.sax.*;
import javax.xml.parsers.*;
import org.apache.soap.util.xml.*;

public class ApacheSoapId {

    public String ServiceId () throws Exception {
        return "This is Apache SOAP";
    }
}
```

The following deployment descriptor is used to make the Web Service available to consumers.

```
<isd:service xmlns:isd="http://xml.apache.org/xml-soap/deployment"
    id="apache-soap-id">
  <isd:provider type="java"
                scope="Application"
                methods="ServiceId">
            <isd:java class="samples.javaid.ApacheSoapId"/>
  </isd:provider>
  <isd:faultListener>org.apache.soap.server.DOMFaultListener

</isd:faultListener>

</isd:service>
```

Then, to deploy the service, execute the following command with the environment set up for Apache *SOAP*. Remember the environment must have all the proper *SOAP* libraries in CLASSPATH to execute the following command.

```
java org.apache.soap.server.ServiceManagerClient
    http://localhost:8080/soap/servlet/rpcrouter deploy
    DeploymentDescriptor.xml
```

Both *.NET* and *Axis* have tools that allow you to create proxies to call other Web Services. Creating a proxy saves you time because it generates a lot of standard code that you would need to write otherwise. Proxies are generated on what appears in a WSDL file. To call this *SOAP* service easily from both *.NET* and *Axis*, you need to create the appropriate WSDL file.

To generate WSDL for an Apache *SOAP* service, you need to use the *Java2WSDL* tool found in the Apache *Axis* distribution. Even though you may be committed to using Apache *SOAP* for legacy reasons, you need to use the *Java2WSDL* tool in *Axis* to generate WSDL for your Web Services.

Creating the WSDL is a little tricky. First you need to compile the Web Service code with an environment set up for Apache *SOAP*. Once it's compiled, you need to open another DOS prompt with an environment set up for *Axis* because you'll need the classes from that package to create the WSDL. From that DOS window, execute the following command.

```
java org.apache.axis.wsdl.Java2WSDL -o GetId.wsdl
-l"http://localhost:8080/soap/servlet/rpcrouter"
-n"urn:apache-soap-id" ApacheSoapId
```

The -o switch specifies the output file, which in this case is GetId.wsdl. The -l specifies the location of the Web Service, which is the URL included. The -n identifies which Web Services to call. The final option to specify is the name of the class.

If this command fails for you, open the Web Service in your editor and comment out the package statement. Then recompile it within the environment set up for Apache SOAP. The package statement appears to confuse the Java2WSDL tool when it is trying to generate the namespace statements.

Figure 9.4 shows the output of the *Java2WSDL* tool in *Internet Explorer*.

The WSDL created by the tool is standard and both *.NET* and *Axis* proxy generators will be able to use the information in the file to create the appropriate proxies.

Apache *Axis* Web Service

After looking at the Apache *SOAP* and Microsoft *.NET* examples, you'll probably realize that it is easy to create a Web Service with *Axis*. Consider the following code.

```
public class ApacheAxisId {

public String ServiceId() throws Exception {
```

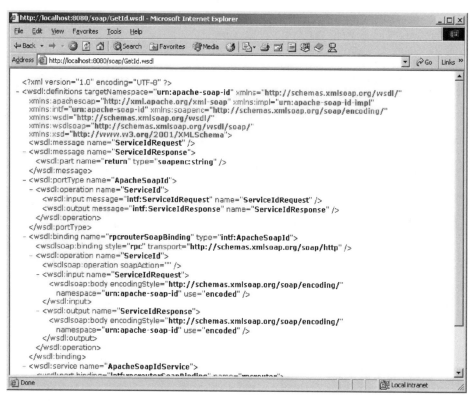

FIGURE 9.4   The WSDL created by the `Java2WSDL` tool for the Apache *SOAP* example Web Service.

```
        return("This is Apache Axis");
    }


}
```

It's amazing that this little snippet of code creates a Web Service. You simply need to drop the code in *Axis'* webapps directory and then point your browser to the following URL to compile the code into the appropriate class file:
http://localhost:8080/axis/ApacheAxisId.jws

By adding "wsdl" to the query string, the Web Service displays the WSDL needed to create a proxy. This Web Service provides the WSDL shown in Figure 9.5.

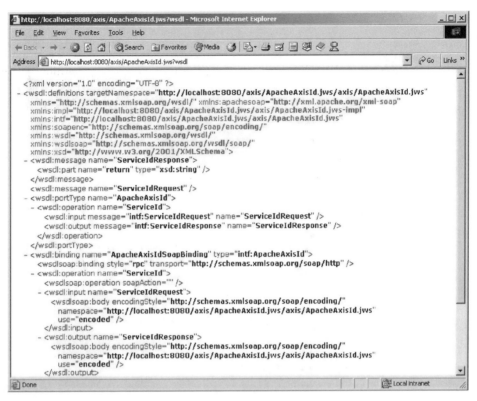

FIGURE 9.5 The WSDL output of the *Axis* example Web Service.

The proxy generator in .NET doesn't always pick up port numbers in URLs. If this happens, you'll receive an HTTP error number 405 indicating that you do not have permission to access the Web Service. To work around this problem, copy the WSDL file to your local system and use this local file to generate the proxy. This somehow fixes the problem.

TESTING WEB SERVICES

Now that you have three Web Services that represent the different environments covered in this book, you need to test them in order to create consumers. The first part of this involves setting up the environment for *Tomcat* such that both *Axis* and *SOAP* Web Services are able to execute simultaneously.

Once the environment is set up correctly, *.NET WebService Studio*, which is provided as a free download from Microsoft on the *www.gotdotnet.com* Web site, allows you to easily test any Web Service that generates WSDL because it builds a proxy and a client dynamically without having to write and compile a program.

Environment Setup

The trick with executing *SOAP* and *Axis* simultaneously is to set up the CLASSPATH and other environmental variables so that all the necessary libraries are available. The CD-ROM for this book has the MSDOS batch file (Example 9-5.bat) the author used to set up the environment for the examples in this chapter.

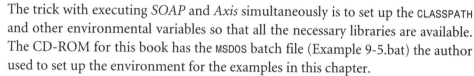

The batch file code looks like the following.

```
set PATH=%PATH%;c:\J2SDK_FORTE\jdk1.4.0\bin;
set JAVA_HOME=c:\J2SDK_FORTE\jdk1.4.0
set CLASSPATH=
    c:\xmlapache\tomcat\webapps\axis\WEB-INF\lib\axis.jar;
    c:\xmlapache\tomcat\webapps\axis\WEB-INF\lib\commons-
        logging.jar;
    c:\xmlapache\tomcat\webapps\axis\WEB-INF\lib\jaxrpc.jar;
    c:\xmlapache\tomcat\webapps\axis\WEB-INF\lib\saaj.jar;
    c:\xmlapache\tomcat\webapps\axis\WEB-INF\lib\log4j-
        1.2.4.jar;
    c:\xmlapache\tomcat\webapps\axis\WEB-INF\lib\tt-
        bytecode.jar;
    c:\xmlapache\tomcat\webapps\axis\WEB-INF\lib\xerces.jar;
    c:\xmlapache\tomcat\webapps\axis\WEB-INF\lib\wsdl4j.jar;
    c:\xmlapache\axis-1_0\;
    c:\xmlapache\javamail\mail.jar;
    c:\xmlapache\jaf\activation.jar;
    c:\xmlapache\soap;
    c:\xmlapache\soap\lib\soap.jar;
    c:\xmlapache\tomcat\lib\servlet.jar;
```

If you look closely at the batch file, you'll notice that it includes all the libraries that *SOAP* and *Axis* need to execute properly. If you can start *Tomcat* and then access both the *SOAP* and *Axis* admin pages, your environmental setup is correct.

Using *.NET WebService Studio*

.NET WebService Studio provides an easy means of testing the compatibility of Web Services with *Dot Net*. By utilizing *.NET WebService Studio*, you'll save time because

you'll know that your Web Services are working properly without having to write and compile a program.

Begin by downloading the *.NET WebService Studio* from *www.gotdotnet.com*. Appendix B contains the information about where to download the file. It is simply an executable, so all you need to do is double-click on the `.exe` file once it's extracted.

Click the "Browse" button and browse to the WSDL file for the *ApacheAxixId*-service.

Remember the port problem mentioned in the tip earlier, so if your Web Service resides on a port greater than 80 you may wish to browse to a local file rather than one at a URL. Click on "Get." The application examines the WSDL file and compiles a proxy. This allows you to call the Web Service without writing a client. Once it considers the WSDL, it brings you to a screen that shows available methods. Select the method you want to test by highlighting it, and then select the parameter that gets passed (if any), and enter a value. Clicking "Invoke" causes *WebServiceStudio* to call the appropriate Web Service and print out the reply. Results of such a test are shown in Figure 9.6.

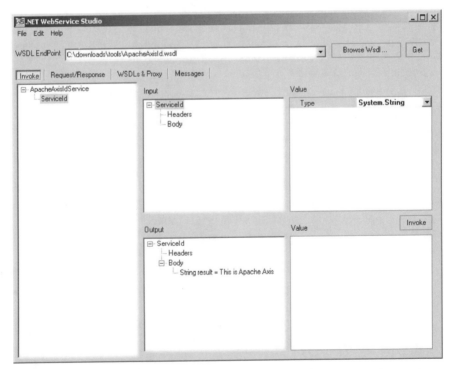

FIGURE 9.6 Notice the output window of the *WebServiceStudio* for the results of calling the *Axis* Web Service.

The "Request and Response" tab allows you to see the *SOAP* messages that are sent and received by the *Axis* Web Service. An example of this is shown in Figure 9.7. This can be very useful during a debugging process where you may be trying to determine why you cannot invoke a particular service.

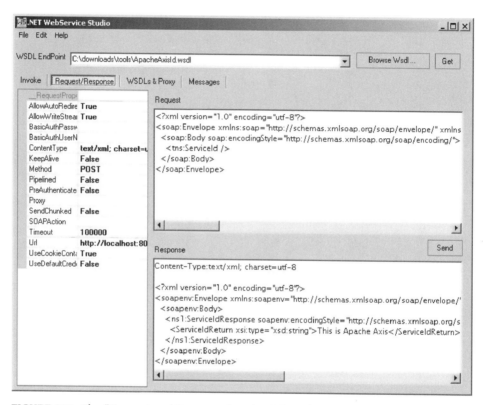

FIGURE 9.7   The "Request and Response" to the *Axis* Web Service in *WebServiceStudio*.

The "WSDL and Proxy" tab allows you to view WSDL file you pointed to. The code generated for the proxy is shown in Figure 9.8.

WebServiceStudio provides you with a quick means of developing a client for testing your Web Services. As long as WSDL is available, *.NET WebService Studio* can test both *.NET* and Apache Web Services.

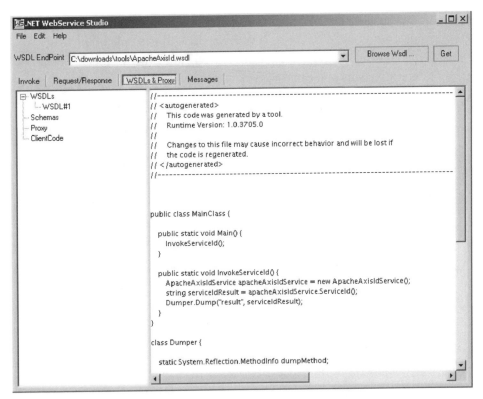

FIGURE 9.8   The client and proxy code shown in *WebServiceStudio*.

CROSS-PLATFORM CONSUMERS

Web Services are most useful when then can bridge the gap between platforms. Traditionally it has been somewhat difficult for applications running on Solaris or another UNIX platform to communicate with a *Windows* application without a large infrastructure such as CORBA in place. Sometimes even communicating with a database can be difficult in this situation.

Web Services provide hope for simple integration between platforms. With just a Web server and some software, applications can bridge the gap between platforms fairly easily. This section covers consumers that call not only *.NET* but also Java Web Services at the same time.

.NET GUI Consumer

.NET comes with several tools that help you consume Web Services. The main caveat rests with the fact that *.NET* requires WSDL to make a proxy that communicates with the service. Ensuring that your project has the ability to generate WSDL greatly increases the chances of compatibility with Web Services under *.NET*.

Remember that it is entirely possible that you can receive WSDL that is perfectly valid but still isn't compatible with .NET. Therefore, it is important for you to test early and often for compatibility with .NET while creating a cross-platform Web Services project.

Create proxies for each of the Web Services shown in this chapter in *Visual Studio.NET* or with the WSDL tool. For further information on how to create proxies for *.NET*, refer to Chapter 6.

With the Java Web Services you'll find that *.NET* will give the Java Web Services names that are hard to use. For example, the name will mimic a package name in Java that might look like this: `samples.simplestock.SimpleStockService` (which is how you would need to use the name in C#). You can rename these class names in the "Solution Explorer" window of *Visual Studio.NET* and, thus, make your coding in C# clearer. For the following example, the Web references for the Java services were renamed to `SoapId` and `AxisId`. Figure 9.9 shows the renamed Web Services in the "Solutions Explorer" window.

Now create a C# windows application in *Visual Studio.NET*. Drag and drop one text box and three buttons onto the form. Label the buttons according to which Web Service they call. The following code snippet highlights the code behind each button.

Because both examples use proxies to call the *.NET* and Java Web Services, the code to call each service is very similar. The first step is to define a String that can accept the return value of the method. The next step is to create an object of the class that utilizes the Web Service. Then the method is called with the new object and the result is put into the String defined earlier, and then displayed in the text box. This code is repeated for all three buttons with the only difference being the objects called.

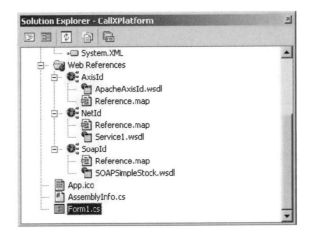

FIGURE 9.9 "Solution Explorer" in *Visual Studio.NET* displaying the renamed Web References.

```csharp
private void button1_Click(object sender, System.EventArgs e)
{
   String returnValue;
   NetId.Service1 idNum1 = new NetId.Service1();
   returnValue = idNum1.ServiceId();
   textBox1.Text = returnValue;
}

private void button2_Click(object sender, System.EventArgs e)
{
   String returnValue;
   SoapId.ApacheSoapIdService idNum2 =
   new SoapId.ApacheSoapIdService();
   returnValue = idNum2.ServiceId();
   textBox1.Text = returnValue;
}

private void button3_Click(object sender, System.EventArgs e)
{
   String returnValue;
   AxisId.ApacheAxisIdService idNum3 =
   new AxisId.ApacheAxisIdService();
   returnValue = idNum3.ServiceId();
   textBox1.Text = returnValue;
}
```

Once you compile and execute your code, the running program looks like Figure 9.10.

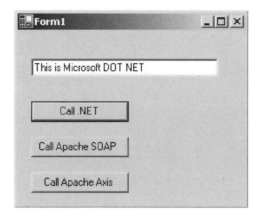

FIGURE 9.10   The `CallXPlatform` program calling a *.NET* Web Service.

Notice that the first button has already been clicked and the results of calling the *.NET* Web Service are shown in the text box.

Clicking on the second button calls the Apache *SOAP* Web Service. Figure 9.11 shows that the *.NET* client in this case was successfully able to contact the service.

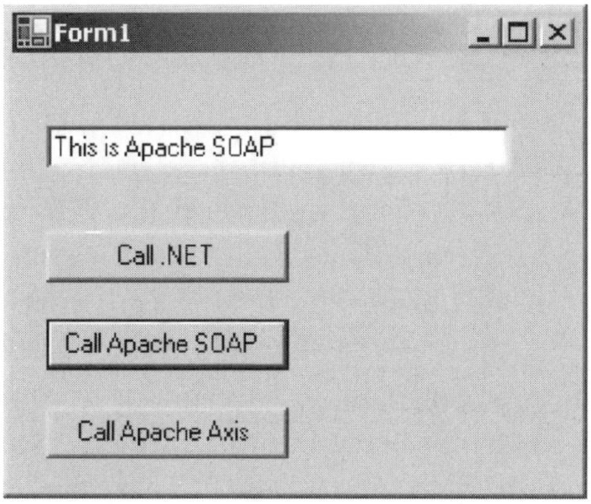

FIGURE 9.11   The `CallXPlatform` program calling an Apache *SOAP* Web Service.

The third button calls the Apache *Axis* service, and also displays the results. Figure 9.12 shows the `CallXPlatform` program calling the Apache *Axis* Web Service.

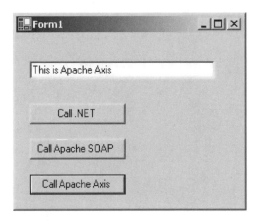

FIGURE 9.12   The `CallXPlatform` program calling an Apache *Axis* Web Service.

Think about this example. With very little work you easily integrated a piece of software written in C# with services created in both *.NET* and Java. This gives you a great deal of flexibility because you can pick and choose which task you want each technology to accomplish. For example, *Visual Studio.NET* comes with a lot of tools that make creating *Windows* programs fast and easy—much easier, in fact, than using a Java *Interactive Development Environment* (IDE) such as *Sun One Studio* or a different Java IDE under *Windows*. By choosing the Microsoft technology, you are able to pick the tool that best fits the application. In another situation, it may make sense to use a Java client on a UNIX system because that's really your only choice. Microsoft claims that eventually *.NET* will be available under Linux and possibly other flavors of UNIX, but for now it is easy to choose which technology is best for the platform with which you are working.

For example, the majority of people in the workplace have a PC on their desktop, but many large corporations have Solaris or some other UNIX platform running on large servers. By being able to utilize Web Services on either platform, you can either write an application for the PC, have Web pages that communicate with the Web Service that can be hosted on either platform, or have applications running on the large UNIX servers that also integrate with the Web Services.

Java Program

As you've seen in previous chapters, it is not necessary to create a proxy for a Java program to call any Web Service, but it does make coding the call much easier and succinct. Chapter 8 discusses how to create proxies for Java. Here it's just a matter of pointing the *WSDL2Java* tool in *Axis* to the appropriate URL that contains the WSDL.

The following example creates a java proxy for for the *.NET* Web Service.

```
java org.apache.axis.wsdl.WSDL2Java
http://localhost/XPlatform/MSNETID/Service1.asmx?wsdl
```

Then, to create the proxy for the Apache *SOAP* service, you need the URL of the WSDL file. In the following example, the WSDL file sits within the *SOAP* directory of *Tomcat*.

```
java org.apache.axis.wsdl.WSDL2Java
http://localhost:8080/soap/GetId.wsdl
```

Next, to call the *Axis* WSDL, simply send the proxy tool the URL of WSDL. Remember that *Axis* is just like *.NET* in that by putting WSDL in the query string you'll get the appropriate code.

```
java org.apache.axis.wsdl.WSDL2Java
http://localhost:8080/axis/ApacheAxisId.jws?wsdl
```

When these proxies are created, they all have different package names based on the namespaces the proxy tool encounters. You'll notice that the following Java client imports several packages that have names based on the namespaces found in the WSDL. Remember that to utilize these packages you'll need to change into the appropriate directory and compile the *.java* files.

The following example defines the class GetAllTypes, which simply contains a main method that utilizes all the proxies created previously to call all the different Web Services. Then, code similar to the examples in Chapter 8 calls each of the example Web Services.

```
import localhost.*;
import apache_soap_id.*;
import com.advocatemedia.www.*;

public class GetAllTypes {
```

```java
public static void main(String [] args) throws
Exception {
//Set up Var for results
 String result;

//Call MS .NET
//The package name comes from the namespace
//designated in the .NET Web Service
Service1Locator myService1 = new Service1Locator();
com.advocatemedia.www.Service1Soap myNET
=myService1.getService1Soap();
result = myNET.serviceId();
System.out.println(result);

result = null;

//Call Apache SOAP
ApacheSoapIdServiceLocator myService2 = new
ApacheSoapIdServiceLocator();
apache_soap_id.ApacheSoapId mySoap =
myService2.getRpcrouter();
result = mySoap.serviceId();
System.out.println(result);

result = null;

//Call Apache Axis
ApacheAxisIdServiceLocator myService3 = new
ApacheAxisIdServiceLocator();
localhost.ApacheAxisId myAxis =
myService3.getApacheAxisId();
result = myAxis.serviceId();
System.out.println(result);
  }
}
```

Figure 9.13 displays the output of the previous example from the DOS prompt. With this example, Java calls all the different types of Web Services covered in this book including *.NET*. With Java being so cross-platform compatible, even a program executing under UNIX or another platform that supports Java could call a Web Service in *.NET*.

FIGURE 9.13 The output of `GetAllTypes` Java program that connects with all three of the Web Services.

THIRD-PARTY TOOLS AND OTHER LANGUAGES

The book focuses on Web Services for Java and C#. The tools necessary to make Web Services for these languages are freely available by downloading tools from Apache or Microsoft. What if you need to support another language or platform?

In some cases, platforms such as a newer IBM mainframe run Linux and, therefore, run Java, *Tomcat*, and Apache Web Services. The source code is available for all Apache products. If a binary isn't available for your platform and a Java compiler is, you may be able to build it for your platform.

If your platform doesn't support Java or you have a legacy system based on something like COBOL, a third-party vendor is probably your only option unless you have a staff who can write the software that supports Web Services. This is a very expensive undertaking because of the time it will take your developers to study the WSDL, *SOAP*, and other standards necessary to follow for compatibility with other implementations.

There are vendors that support such legacy systems such as Cape Clear Software, *www.capeclear.com*, that has a plug-in for COBOL. In addition, other vendors such as Sysinet systems, *www.sysinet.com*, also have support for Web Services in C++. The Apache group plans to have C++ client available in the Fall of 2002.

Another advantage that third-party vendors offer with Java is easy integration with *Interactive Development Environment* (IDE) such as *Forte™* or *Jbuilder™*, but the biggest factor in your decision about whether to use Apache software or a third-party vendor may be support. Although the Apache group does provide support in the form of discussion groups on the Internet, if you are doing something slightly out of the mainstream you may not get the desired response to your question. By going with a third-party vendor, you guarantee that you have access to support at critical moments. It really depends on the project you undertake. If you are doing something very mainstream, Apache is a great choice. If your project involves something a bit uncommon such as utilizing attachments or modifying the *SOAP* header for compatibility with other technologies, a third-party vendor may be a great choice. Very often the choice of technology depends on the company culture and requirements made by management.

By utilizing third-party, open source, and Microsoft-based Web Service tools you can eliminate the need for staging data from other sources. In many cases, data from COBOL and other legacy systems are staged in large databases or text files because the software isn't available to access the systems in other ways. The data stores are then updated on a regular basis but are still almost always out of date. If you are able to put a Web Services layer on top of these legacy systems, you can access data immediately.

Another advantage of integrating all these systems is delaying the purchase of new software and hardware. If all the systems can integrate regardless of OS, you can continue to utilize these systems and have them communicate without large capital investment. Figure 9.14 illustrates the communication that may occur in an enterprise setting.

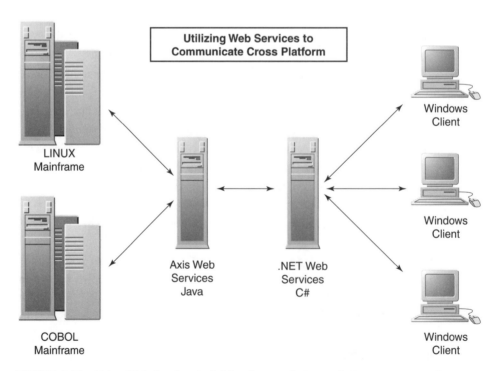

FIGURE 9.14   Using Web Services to bridge the gap between heterogeneous systems.

Note that the C# Web Services focus on serving *Windows* clients and the Java Web Services deal with communicating with the corporate systems. Then, the servers that host the services act as a bridge between Java and C#. Although the flexibility of *Axis* allows your *Windows* clients to contact these Web Services directly, using the C# Web Services allows your *Windows* clients to consume Web Services from the same location. This makes handling the details of Web Services easier for an administrator.

CONCLUSION

The tools that come with both Apache *Axis* and Microsoft *.NET* allow for easy cross-platform integration. The first Web Services release from the Apache group, Apache *SOAP*, didn't have the tools necessary to easily create cross-platform Web Services. Because *SOAP* was out long before *.NET*, the developers were in the dark on what tools were really necessary to work with future Microsoft releases. Many of the tools provided with *Axis* such as the *wsdl2java* tool and the ability for the Web Service to generate its own WSDL allow Java developers to have the same ease of use as the *.NET* community, along with the ability to call C# Web Services.

When Microsoft first started creating and releasing Web Service tools, several third- party Java vendors rushed to the scene with Java Web Service tools. The only open source competition then was Apache *SOAP*, which didn't provide many of the tools needed to easily integrate between Java and *.NET*.

The release of the "Web Services Toolkit" by IBM was the first real hint of easy cross- platform integration that a developer had free access too because the toolkit was available as a free download. Once IBM transferred some of the code from the toolkit to the Apache Group, the *Axis* project began to compete with many of the third-party vendors who rushed to develop tools that integrated with *.NET*. One must wonder what the future of the third-party Java Web Services vendors is?

10

Web Service Security

In This Chapter

- Security Considerations
- Implementing Web Service Security
- Cross-Platform Web Server Security
- The Future of *SOAP* and XML Security

W̶eb Services are easy to implement and use cross platform, but perhaps their biggest weakness is the lack of security built into the *SOAP* and other Web Services standard. The fact that Web Services utilize port 80 and HTTP means that they use a port probably configured on your firewall policy that isn't protected from the outside world. Additionally, the very nature of the Internet—with proxy servers and the overall openness of information makes your Web Services vulnerable. Remember that anyone with a proxy server or a node can view your request or response with relative ease. Consider the *SOAP* for one of the `SimpleStockQuote` examples shown earlier in this book.

```
Content-Type: text/xml;
charset=utf-8 Content-Length: 461
SOAPAction: ""
<?xml version='1.0' encoding='UTF-8'?>
<SOAP-ENV:Envelope xmlns:SOAP-
        ENV="http://schemas.xmlsoap.org/soap/envelope/"
            xmlns:xsi="http://www.w3.org/2001/XMLSchema-instance"
            xmlns:xsd="http://www.w3.org/2001/XMLSchema">
 <SOAP-ENV:Body>
   <ns1:getTestQuote xmlns:ns1="urn:simple-stock-quote"SOAENV:
       encodingStyle="http://schemas.xmlsoap.org/soap/encoding/">
        <symbol xsi:type="xsd:string">C</symbol>
   </ns1:getTestQuote>
 </SOAP-ENV:Body>
</SOAP-ENV:Envelope>
```

Not only is the information in text format rather than a binary format but also the content of the document is surrounded by self-describing XML. Thus, the information is easy to view; you took the time to make it easier for the eavesdropper to understand the type and the content of the information you are transmitting. With a comma-delimited string of information, at least, the eavesdropper has to determine what the data actually represents.

The value of a stock, however, is not the most secretive piece of information. Thus you probably would not need a great deal of security to protect that information. On the other hand, if you handled a consumer's credit card information, you would definitely need to encrypt that information as the request and response made their way across the Web.

Beyond encryption, there are also security concerns you need to consider with Web Services. Such security concerns mainly involve deals the identity of the individual or application accessing your service. Is it whom you expect? Is someone impersonating your regular consumer? Identity isn't something you consider when deploying a HTML form on your Web site, but with Web Services, you're providing functionality that possibly interacts with your backend systems such as your *Customer Relations Management* (CRM) or accounting database at an application level. You need to protect this information.

As with any with any Web project, you need to sit down and consider the amount of security you really need because of the impact on hardware. Once you know your security needs, you can better answer questions related to needed systems.

SECURITY CONSIDERATIONS

Evaluating the amount and focus of security you need is a process that you must consider for any project, not only for a Web Services project. The higher the security, especially with encryption, the greater impact on processing power and memory on the server. This section covers many aspects you need to consider for security purposes.

The Firewall

A firewall is a piece of network hardware or software that regulates the ports and protocols a server accepts or denies. This hardware often sits wherever Internet access comes into a corporation through a T1 line, DSL connection, or some other connection to a provider. It usually allows incoming and outgoing HTTP, telnet, and FTP requests on standard ports; it also determines which internal servers are available to the outside world. Figure 10.1 shows a firewall routing request based on its policies. Notice the role the proxy server plays in these requests. The firewall server allows incoming HTTP and allows outgoing HTTP, telnet, FTP, and nothing else. The proxy server examines each HTTP request, including a Web Service request, and ensures that the request accesses an approved site.

Proxy servers have many more uses, such as caching.

NOTE

Servers that are often involved in internal projects are not available to consumers or hackers from the Internet. Thus, you can relax your concerns about security. This also means your project is probably free from someone looking at your *SOAP* messages through a proxy server. A proxy server exists near the corporation's

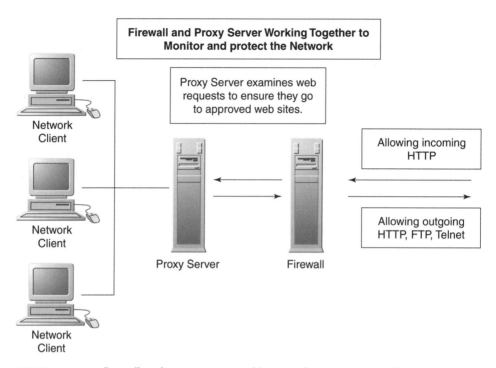

FIGURE 10.1 A firewall and proxy server working together to protect and monitor the network inside a corporation.

firewall (not necessarily *physically* next to it) and ensures that employees are only viewing legitimate Web sites, which include viewing Web Service requests and responses.

There are two layers of security here. Protection from the outside world and not going through a proxy. This doesn't mean that you can't have internal malicious users access your system, but it does mean that you have greater control over these individuals when they do something inappropriate.

The Network

By using network hardware and configuration, you can limit and control access. Within an organization it may be possible to put the application and its consumers within their own subnet, where only those users have access to the application through the subnet, but are still able to communicate with the rest of the network. This requires that your application be internal where you have control over the network setup. Figure 10.2 shows a subnet regulating access to a Web Services application. The switch allows traffic out to the rest of the network, but denies access to anything within the subnet.

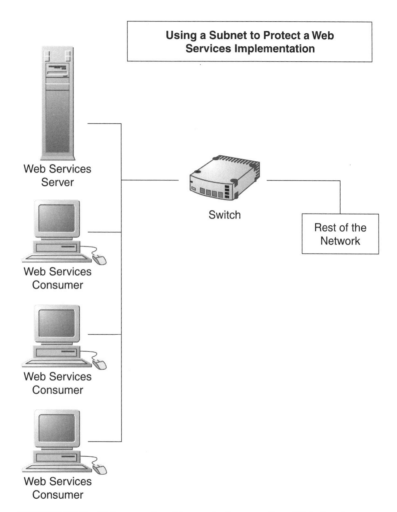

FIGURE 10.2   Using a subnet to control access to a Web Services implementation.

The same concept can be applied to external customers with the configuration of the Web server controlling your Web Services. Web servers allow you to determine which *Internet Protocol* (IP) addresses and domains can access your site. Combining this with configuring the network, you should end up with reasonable security.

This only works, however, if your application is for a unique group of users. You also need to be concerned about Dynamic IP addresses. In this case, users may have a different IP address each time they connect to your service. This is especially true if your user comes through a dial-up *Internet Service Provider* (ISP) rather than through a corporate firewall, but it often happens with DSL and cable too. These are definitely things to consider when trying to use a network or Web server configuration to control access to your services.

Security Focus

The focus of the security you need to implement depends on your application. If you're providing a subscription to a Stock Quote service, you want to ensure that your user transmits some sort of credentials to ensure he is a paying user, but the information, such as a value of a stock, is not confidential information. So your focus in this case is on the identity of the user rather than the data.

In another case, you want to ensure that the user identity and the information transmitted are both protected. This is especially true when dealing with credit card information. Remember that you transmit text-based information in an XML file, and thus someone could easily intercept a *SOAP* document and determine what the information is, as illustrated in the following example.

```
Content-Type: text/xml;
charset=utf-8 Content-Length: 461
SOAPAction: ""
<?xml version='1.0' encoding='UTF-8'?>
<SOAP-ENV:Envelope xmlns:SOAP-
        ENV="http://schemas.xmlsoap.org/soap/envelope/"
            xmlns:xsi="http://www.w3.org/2001/XMLSchema-instance"
            xmlns:xsd="http://www.w3.org/2001/XMLSchema">
 <SOAP-ENV:Body>
   <ns1:purchaseBooks xmlns:ns1="purchasebooks"
        SOAP-
ENV:encodingStyle="http://schemas.xmlsoap.org/soap/encoding/">
        <CreditCard xsi:type="xsd:string">4790-0911-0000-
            0000</CreditCard>
        <Expires xsi:type="xsd:string">10-05</Expires>
        <OrderId xsi:type="xsd:int">32111</OrderId>
    </ns1:puchaseBooks>
```

```
</SOAP-ENV:Body>
</SOAP-ENV:Envelope>
```

One of the benefits of XML for hackers is that it is self-describing. In other words, when using XML not only do you provide the information in an easy-to-read format but also you insert tags that tell the hackers what the data is! In a case like this, you want to ensure that the data is encrypted and that you send this information to the proper person. Thus, identity and data need to be protected in this example.

When dealing with information such as stock quotes or weather, your security focus centers on identifying the individual. In other cases you may want to protect the data from prying eyes with some sort of encryption, especially with credit card or banking information. You also want to be sure that the identity of the consumer is confirmed.

Hardware and Software Availability

Security often adds to the load on a server, whether it is authentication, Secure Socket Layer, or another form of security. Security that requires encryption and decryption tends to place heavy loads on servers. Therefore, you need to either prepare your system or purchase another system to handle security.

Beyond hardware, you need to ensure that your Web server or other software handles the security that you need. For example, the version of IIS on *Windows 2000 Pro* does not handle SSL, so if you were using that software, you would need to consider upgrading your system to one that handles SSL properly. This can be done by upgrading to a server version of *Windows* such as *Windows 2000* server or *Windows XP* server.

If your budget doesn't allow for the purchase of a new system or systems, you need to consider a security option that fits within your current system. Ensuring that your security needs fit your software and hardware requirements is an important consideration.

IMPLEMENTING WEB SERVICE SECURITY

After considering the previous section, you should probably possess a reasonable understanding of your security needs. This section of the chapter provides examples for implementing some of the security requirements covered in the previous section.

This section begins by looking at identifying the user through authentication and looking at hostnames and IP addresses. Then the focus centers on utilizing SSL as a means of encrypting the connection between Web Service and consumer.

Providing an `ID`

This is a simple but effective means of securing your Web Service. It simply involves creating a unique `ID` for each consumer who accesses your service. This can be a simple implementation in which this unique `ID` is generated by a column in a database table or it can be more complicated by using *Globally Unique ID* (GUID). (For more information on GUIDs, be sure to see *www.guid.org*.)

Any client-side authorization requires that you communicate with your consumers so that they can easily obtain an `ID`. *When you are designing your services, you want to make sure you plan for a means of communication with your consumers so they can obtain this unique* `ID` *quickly and easily. It is best if this can be an automated process.*

Consider the following *Axis* Web Service.

```
public class SimpleStockExample {

public float getTestQuote(String symbol, String ID)
    throws Exception {

  //Have code call a db or other
  //call to compare ids.
  if (ID.equals("a123jj5!") {

    if ( symbol.equals("C") ) {
        return( (float) 55.95 );
    } else {
        return( (float) -1);
    }

  } else {
        return( (float) -2);
  }
```

```
    }

  }
```

This is a rather simple example but it illustrates the point. ID is a String variable that contains a unique identifier. The method checks to ensure that the identification is present and correct.

If you look at the previous example code closely, you'll notice that the method acts differently depending on whether symbol or the ID is passed the wrong value. This is important to your users so that they know whether it's a problem with the identification or with the method itself. This also gives them a way to test their connection to the Web Service before they receive their ID from you.

This example is overly simplistic; you need to have a complex mechanism in place to create unique IDs for your users. For example, you may have a large CRM database that generates unique user IDs for you. By putting in some SQL where the code currently checks for a particular string you would have a powerful way of validating a user.

Using the Web Server for Authentication

One of the basic functions of a Web server involves securing Web pages, CGI scripts, Active Server Pages, and other server-side code from unauthorized visitors. Because Web Services utilize port 80 and usually integrate with a technology close to a Web server such as *.NET* or *Tomcat*, configuring a Web server to protect a Web Service is usually a simple task. This section focuses on the steps necessary to use a Web server for this form of protection.

Note that anytime you involve the Web server for authentication, you're adding another layer that will definitely affect the performance of the Web Service.

Security with *Tomcat* and *Axis*

Axis allows you to set some security within the deployment descriptor. The following XML file deploys an *Axis* Web Service to look for a text file containing usernames and passwords. Notice that there are extra parameter tags in this descriptor that define the `allowedMethods` and the `allowedRoles`. These parameters explicitly define the methods accessible and who may access them.

Then the handler elements tell *Axis* to load the `SimpleAuthenticationHandler` and `SimpleAuthorization` classes into memory so the security handling can occur.

```
<deployment name="myDeployment"
            xmlns="http://xml.apache.org/axis/wsdd/"
            xmlns:java=


    "http://xml.apache.org/axis/wsdd/providers/java">

<service name="SimpleStockExample"
        provider="java:RPC">
<parameter name="className"
           value="SimpleStockExample"/>
<parameter name="allowedMethods"
           value="getTestQuote"/>
<parameter name="allowedRoles"
           value="brianhochgurtel davidp"/>

<requestFlow name="checks">
  <handler type=
      "java:org.apache.axis.handlers.SimpleAuthenticationHandler"/>
  <handler type=
      "java:org.apache.axis.handlers.SimpleAuthorizationHandler"/>
</requestFlow>
</service>

</deployment>
```

A file is then created, called *users.lst*, which belongs in the root directory of the *Axis* webapps directory under *Tomcat*, which looks like the following.

```
brianhochgurtel somepassword
davidp
```

The user `brianhochgurtel` is required to have a password whereas the user `davidp` is only required to submit his username.

When a client connects, the request must pass the required credentials to the server. The username and password (if necessary) simply need to be passed on as part of the request in the consumer. This can be done either as part of a command line argument or in the code.

Then, in the consumer, you add the username and password just like you add the name of the service and method you are calling. Note that this code snippet assumes that you have the value of the username and password in the variable's `user` and `passwd`.

```
        //Remember that the call object contains all the
    //information that gets passed to the service.
    Call    call    = (Call) service.createCall();

    call.setTargetEndpointAddress( url );
    call.setOperationName
    ( new QName("SimpleStockExample", "getTestQuote") );

    //Other parameters need to utilize the Web Service would
    //appear here.

    //set the username and password in the call
    call.setUsername( user );
    call.setPassword( passwd );
```

If you don't call the protected Web Service with the username and password, you'll get a `405 Error` stating that you are not authorized.

Using *Windows*-Integrated Authentication

When compiling a Web Service generated in *Visual Studio.NET*, a `web.config file` is also created. It is an XML file containing configuration directives that normally aren't available in the code. The following XML is an example.

```
<?xml version="1.0" encoding="utf-8" ?>
<configuration>

  <system.web>


    <compilation
        defaultLanguage="c#"
        debug="true"
    />
```

```
<customErrors
mode="RemoteOnly"
/>


<authentication mode="Windows" />


<trace
    enabled="false"
    requestLimit="10"
    pageOutput="false"
    traceMode="SortByTime"
      localOnly="true"
/>

<sessionState
        mode="InProc"
        stateConnectionString="tcpip=127.0.0.1:42424"
        sqlConnectionString="data source=127.0.0.1;user
        id=sa;password="
        cookieless="false"
        timeout="20"
/>

<globalization
        requestEncoding="utf-8"
        responseEncoding="utf-8"
  />

  </system.web>

</configuration>
```

If you look closely at the XML you'll find that it sets the language that compiles the service, the authentication mode, whether trace is enabled, and other information. It is here that you can add some XML that either enables or disables a user's access to a particular Web Service.

By adding the following XML to the web.config file, you begin to regulate the users who access your Web Service.

```
<system.web>
 <authentication mode="Windows" />
</system.web>


<location path="Service1.asmx">
    <system.web>
        <authorization>
          <deny users="*"/>
        </authorization>
      </system.web>
  </location>
```

This XML forces the authentication to use *Windows* and specifies that Ser-vice1.asmx is the protected file. In this case, all users are denied, which might be a handy way to prevent users from accessing your service while you are servicing it. Other possible options look like the following.

```
        <allow users="Administrator"/>
  <allow users="Brian Hochgurtel"/>
```

The entire web.config file now looks like this:

```
<system.web>
   <authentication mode="Windows" />
</system.web>


<location path="Service1.asmx">
  <system.web>
    <authorization>
        <allow users="Administrator"/>
        <allow users="Brian Hochgurtel"/>
    </authorization>
  </system.web>
</location>


<system.web>


<compilation
```

```
        defaultLanguage="c#"
        debug="true"
/>

<customErrors
mode="RemoteOnly"
/>

<trace
    enabled="false"
    requestLimit="10"
    pageOutput="false"
    traceMode="SortByTime"
      localOnly="true"
/>

<sessionState
        mode="InProc"
        stateConnectionString="tcpip=127.0.0.1:42424"
        sqlConnectionString="data source=127.0.0.1;user
        id=sa;password="
        cookieless="false"
        timeout="20"
/>

<globalization
        requestEncoding="utf-8"
        responseEncoding="utf-8"
/>

</system.web>

</configuration>
```

Then you need to go into IIS and select the file you wish to secure, in this case service1.asm, and right-click on it and select "Properties." Select the "File Security" tab. Figure 10.3 shows the dialogue box's security tab.

The first section of this dialogue deals with enabling anonymous access and authentication control. Click on the "Edit" button and another dialogue box appears, as shown in Figure 10.4.

Disable anonymous access and basic authentication and click on the checkbox for "Integrated *Windows* Authentication." This forces *Windows* authentication on

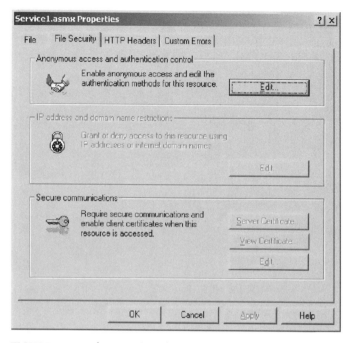

FIGURE 10.3   The security tab in Internet *Information Server*.

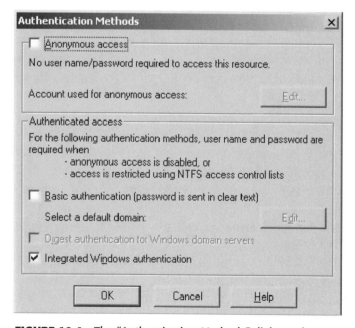

FIGURE 10.4   The "Authentication Methods" dialogue box.

the Web Service. The next time you try to access the Web Service via a browser, Web page, or other consumer, a dialogue box will appear to ask you to log in, as shown in Figure 10.5.

FIGURE 10.5 The *Windows* authentication dialogue box.

Using the *Windows*-integrated login works well if all your consumers use Microsoft *Windows*. If anyone uses a different platform, you'll have trouble using this form of authentication. However, if your users are exclusively on *Windows*, this method prevents people on other platforms from accessing your service.

If all your consumers utilize Windows *but you need to use Java, it is possible to use* Tomcat *(or another Java container) and IIS together. This way you may use* Apache Axis *under* Windows *and use IIS as a Web server. For more information, be sure to see the following URL:*
http://jakarta.apache.org/tomcat/tomcat-3.3-doc/tomcat-iis-howto.html.

Using IIS Basic Authentication

Windows-Integrated Authentication requires that you execute your application on a machine running *Windows* because the security is so tightly integrated with the operating system. This obviously won't work in a cross-platform environment. Thus, if you still need to utilize security based on usernames and passwords on a particular machine or domain, IIS Basic Authentication is the answer. It still uses usernames and passwords from Windows but uses an authentication scheme that isn't as closely tied to the operating system.

Go back to the "Security" tab and select "Basic Authetication," as shown in Figure 10.6.

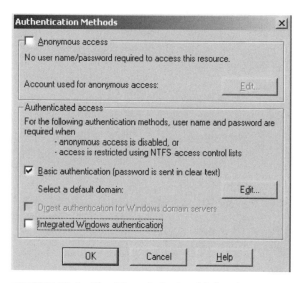

FIGURE 10.6 The "Security" tab with "Basic Authentication" chosen.

If you need to specify a domain for the authentication, click on the "Edit" button next to the "Basic Authentication" and select a local domain on your network. Figure 10.7 shows this dialogue. Note that on the author's machine, the only domain available is "Computers Near Me."

When you try to access this server with a browser, you'll see the prompt for the username and password. Figure 10.8 displays the "Basic Authentication" dialogue box.

Remember that you didn't change the code of the Web Service to integrate it with security under IIS. You just changed the configuration of the Web server.

Next, to access the Web Service with this authentication, use the following code within the client. By setting idNum1 object's attribute Credentials to a new Network-Credential object, the call to the Web Service now contains the information necessary to authorize itself during the transaction.

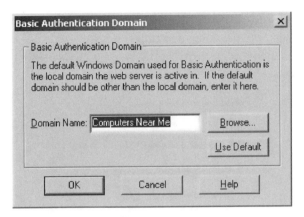

FIGURE 10.7 Selecting a domain name for users that utilize Basic Authentication to access the Web Services.

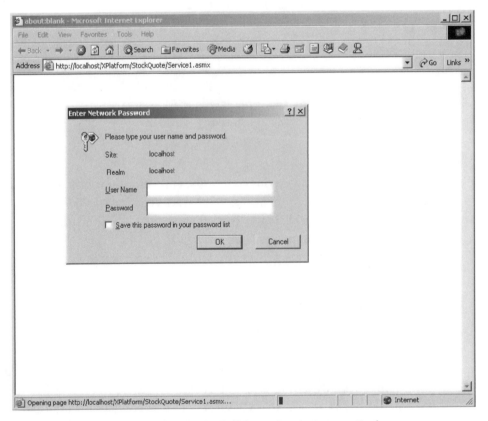

FIGURE 10.8 The "Basic Authentication" dialogue box in *Internet Explorer*.

```
private void button1_Click(object sender, System.EventArgs e)
  {
  String returnValue;
  NetId.Service1 idNum1 = new NetId.Service1();
  idNum1.Credentials =
  new System.Net.NetworkCredential("Some User","password");
  returnValue = idNum1.ServiceId();
  textBox1.Text = returnValue;
  }
```

It is possible to add a third parameter to the `NetworkCreditial` object that specifies a *Windows* domain.

In addition to the previous code, when you add a Web reference in *Visual Studio*, the dialogue prompts you for a username and password so it can access the WSDL, as shown in Figure 10.9.

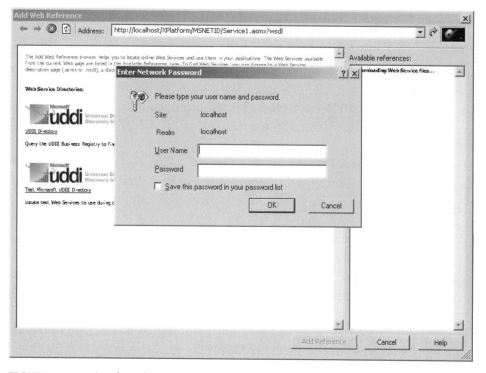

FIGURE 10.9 *Visual Studio.NET* asks for your username and password to access the WSDL of a secured Web Service.

If you need to use the WSDL tool from the command line, the following example demonstrates using it with the /username and /password options, so you may access the WSDL.

```
wsdl /language:cs /username:brianhochgurtel /password:somepassword
/namespace:www.advocatemedia.com
http://localhost/XPlatform/MSNETID/Service1.asmx?wsdl
```

This will authenticate the WSDL tool's request for the WSDL file from the appropriate Web Service.

Looking at IP Addresses and Domains

In this section, we'll focus on the Apache Web server's ability to restrict access to particular files. *Tomcat* does not have this ability, but Apache allows you to configure it so it can send and receive requests to *Tomcat*. Remember that *Tomcat* is a container that allows you to run Java code and not much else whereas Apache is a fully functioning Web server that has the ability to assign different security measures to a particular directory.

Tomcat and Apache communicate through the *Apache JServ Protocol* (AJP), which is similar in some ways to telnet. This connection is open between *Tomcat* and Apache and is a high-speed data connection for them to transfer requests and responses between each other. This is much faster than simply having Apache redirect requests to *Tomcat* through HTTP.

For information on how to configure Apache with Tomcat, *please see the following URL on the Apache Group's Web site:* http://jakarta.apache.org/tomcat-old/jakarta-tomcat/src/doc/tomcat-apache-howto.html

Consider that the SimpleStockExample *Axis* Web Service is normally accessible through the following URL: *http://localhost:8080/axis/services/SimpleStockExample*. Remember that port 8080 is the port that *Tomcat* utilizes, but once it is configured to work with Apache you will no longer need to specify the port. Now the URL is the following: *http://localhost/axis/services/SimpleStockExample*.

To configure security based on IP address or domain, you need to edit the Apache Web server's *httpd.conf file* for the whole system, which includes all URLs the server handles with the following code.

```
      Order Deny, Allow
Deny from all
Allow from advocatemedia.com charlesriver.com
```

The `Order` statement tells the server to handle the `Deny` command first and then the `Allow`. `Deny from all` tells Apache to reject all connections unless the configuration file has any `Allow` statements like the next line. `Allow` then sets up the server to allow anyone connecting from the domain `advocatemedia.com` and `charlesriver.com`.

If you wish to protect one particular directory, you need to use the directory element surrounding this directive. Consider the following.

```
<Directory /usr/local/tomcat/webapps/axis>
        Order Deny, Allow
  Deny from all
        Allow from advocatemedia.com charlesriver.com
</Directory>
```

Now everything under the access directory is protected except from users from the two specified domains. IP addresses can be specified instead. Thus, if you want to deny someone at a particular IP address you simply add the following to the *conf* file.

```
deny 216.183.123.21
```

When basing the security on IP addresses, ensure that your consumers possess static IP addresses. Many Networks use Dynamic Host Configuration Protocol (DHCP), *which can assign IP addresses to a system every time a reboot occurs. When this occurs, your consumer will no longer be able to access the Web Service because its new IP address will not be in the configuration file.*

Unfortunately, IIS under Windows XP Pro, Windows 2000, *and* Windows NT *does not allow you to deny users access by IP address. You need the server versions to do this. For information on how to configure IIS under one of the server versions, see the following URL:* http://msdn.microsoft.com/library/default.asp?url=/library/en-us/dnservice/html/service09052001.asp.

Secure Socket Layer

Secure Socket Layer (SSL) encrypts transmission of requests and responses between a Web server and a client. When the client first communicates with the server, a certificate is exchanged between the two and a key remains on the server. Both the certificate and the key contain information to allow the encryption and decryption of requests and responses to the server. Rather than using port 80 like most requests to a Web server, SSL utilizes port 443.For you to take advantage of this security, you must configure Apache to proxy requests to *Tomcat* via the *Apache JServ Protocol* (AJP) because *Tomcat* has no means of utilizing SSL on its own.

Properly installing SSL on a Web server involves purchasing a certificate from a *Certificate Authority* (CA) such as Verisign. Anyone can actually generate a certificate with various software packages, but a CA verifies names and addresses and other information related to the owners of the certificate and actually provides some insurance for the transactions that occur.

Unless you are using the server version of *Windows*, SSL is only available to you using the Apache Web server. You can download a binary version of Apache that is SSL enabled from *http://www.modssl.org/contrib/ftp/contrib/*. Look for a version that includes Apache, *mod_ssl*, and *openssl*. *Openssl* is software that allows you to create a key, a certificate-signing request, and a test certificate.

NOTE

SSL support for IIS exists only for the server versions of Microsoft operating systems, such as Windows 2000 *server, which this book does not support. If you are interested in setting up SSL with Web Services, see the following URL at Microsoft's support site:*
http://support.Microsoft.com/default.aspx?scid=kb;EN-US;q307267.

Creating a Certificate and Key

For testing purposes, it is useful to generate a self-signing certificate. The following batch file examples combine all the necessary `openssl` commands to generate the key and signing request, and actually sign the certificate as well. The inspiration for this batch file comes from the instructions that come with the `openssl` download for Apache with SSL enabled and with the `openssl` executable.

The first step specifies which `config` file `openssl` should utilize when creating the key and the certificate-signing request. A `.conf` file for `openssl` should come with the distribution. When the first line executes, it will ask you questions about your job and location; most of this information is used to create a seed to create a random encryption key. The only question that you need to answer truthfully is for the "Common Name." This should be the URL of the Web Site that you wish to enable SSL on. If it isn't the same as the Web site, a user's browser will throw a warning with each visit to the site. Remember the password that you create in this step because the second step requires it.

The second step removes the password for your key and signing request. This makes it easier for you when creating certificates and requests because it's one less thing to remember. The third step actually signs the X509 certificate. This is usually the step that you let Verisign or another CSA do. When in development, however, creating your own key saves time and allows you to test under SSL. The last step takes the certificate and creates a form of certificate that *Internet Explorer* can place in its certificate store.

```
openssl req -config openssl.cnf -new -out advocatemedia.csr
openssl rsa -in privkey.pem -out advocatemedia.key
openssl x509 -in advocatemedia.csr -out advocatemedia.cert
    -req -signkey advocatemedia.key -days 1000
openssl x509 -in advocatemedia.cert -out
    advocatemedia.der.crt -outform DER
```

Even if you use the test certificate included on the CD-ROM, it's still good for you to see how to create a CSR. If you are ever involved in any sort of Web project, it is likely that you will complete this step.

SSL and Apache

Apache handles SSL very easily if you either compile SSL into your current Apache installation or download a binary version for *Windows* from the Web site *www.openssl.org*. Once you have the download, follow the instructions in the root directory of the extracted files.

The configuration involves telling Apache to listen to port 443, load the SSL module, set commands for openssl, and move a couple of dlls to the WINNT\system directory. If you follow the instructions carefully, you should have an SSL-enabled version of Apache up and running in a few minutes.

Configuring and Testing *Tomcat*

Once your SSL-configured Apache is up and running, you need to configure *Tomcat* to work together with Apache. As mentioned earlier, Apache and *Tomcat* communicate via the AJP protocol, which is a bare-bones telnet-like binary communication stream.

Tomcat creates a lot of configuration files for you to easily integrate with Apache, and it generates these files every time it starts up looking at the different webapps directories you added. In the conf directory of the *Tomcat* distribution, you'll find a conf file named *Apache-Tomcat.conf*. This is a file generated by *Tomcat* when a new webapps directory gets added. Notice that the directories for *Axis* and *SOAP* are already present. The following text is an excerpt of the configuration file.

```
Alias /soap "C:/xmlapache/tomcat/webapps/soap"
<Directory "C:/xmlapache/tomcat/webapps/soap">
    Options Indexes FollowSymLinks
</Directory>
ApJServMount /soap/servlet /soap
<Location "/soap/WEB-INF/">
    AllowOverride None
    deny from all
</Location>
<Directory "C:/xmlapache/tomcat/webapps/soap/WEB-INF/">
    AllowOverride None
    deny from all
</Directory>
<Location "/soap/META-INF/">
    AllowOverride None
    deny from all
</Location>
<Directory "C:/xmlapache/tomcat/webapps/soap/META-INF/">
    AllowOverride None
    deny from all
</Directory>

Alias /axis "C:/xmlapache/tomcat/webapps/axis"
<Directory "C:/xmlapache/tomcat/webapps/axis">
    Options Indexes FollowSymLinks
</Directory>
ApJServMount /axis/servlet /axis
<Location "/axis/WEB-INF/">
    AllowOverride None
    deny from all
</Location>
<Directory "C:/xmlapache/tomcat/webapps/axis/WEB-INF/">
    AllowOverride None
```

```
     deny from all
</Directory>
<Location "/axis/META-INF/">
    AllowOverride None
    deny from all
</Location>
<Directory "C:/xmlapache/tomcat/webapps/axis/META-INF/">
    AllowOverride None
    deny from all
</Directory>
```

The text in the configuration file sets up all the directives Apache needs to proxy requests to it. Once set up correctly, you no longer need to direct your *Tomcat* requests to port 8080. By simply including this conf file in Apache's *httpd.conf file*, the Web server possesses the ability to proxy requests through port 80 to *Tomcat*'s webapps directory.

There are actually several different methods and protocols to configure *Tomcat* to work with Apache. For more information, be sure to see the following URL on the Apache Group's Web site:

http://jakarta.apache.org/tomcat/tomcat-3.2-doc/tomcat-apache-howto.html.

Once you have Tomcat *and Apache working together, you may notice that you have difficulty accessing the Web Services through the links that Apache provides. This has to do with the configuration of* Tomcat *and Apache working together. If you have this problem, you can access the Web Service and the WSDL with a URL such as the following:*

https://localhost/axis/servlet/AxisServlet/xmltoday-delayed-quotes?wsdl

It is a common mistake for people who are new to Apache products to not execute both the Apache Web server and Tomcat *running at the same time. To utilize* Axis *or another server-side Java code and HTTPS,* Tomcat *and Apache must execute simultaneously, but you can configure them to run on separate systems.*

Registering a Certificate

When you participate in E-Commerce over the Internet, your browser handles the security for you because it has a built in mechanism to do so. To use an application with SSL, it is necessary to register that certificate with the client machine—unless the application has a mechanism to handle the details of SSL like your browser. This must be done for both *.NET* Web Services and Apache.

Under *Windows*, registering a client certificate is done using the "Management Console." Start the console by doing a "Start" and "Run," and then type "mmc.exe." Figure 10.10 shows the console when it first loads.

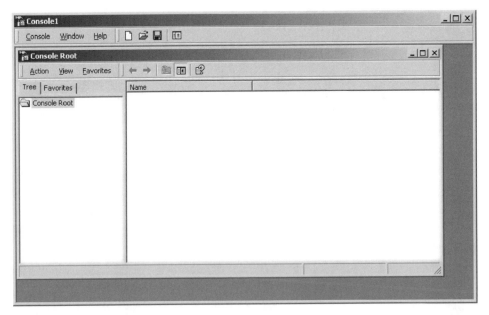

FIGURE 10.10   The "Management Console" on *Windows 2000 Pro* when it first loads.

Click on "Console" and select "Add/Remove Snap In" from the menu. Figure 10.11 shows the new window that appears.

Now click on "Add" and from the list in the window select "Certificates" by double-clicking on the certificate option. Figure 10.12 shows the list of certificates from which you can choose.

A window will pop up and ask who you want to manage the certificates. Select "My User Account" and the console will now look like Figure 10.13.

Select "Trusted Root Authority" and a list of all certificates installed for you on the client side will be shown. From the "Action" menu select "All Tasks" and then

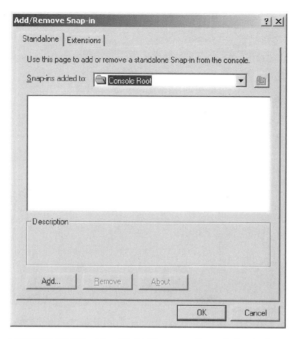

FIGURE 10.11 The "Add/Remove Snap In" window.

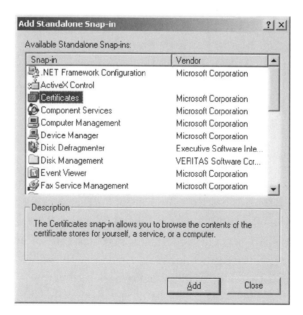

FIGURE 10.12 The window in the management
console where you select certificates to manage.

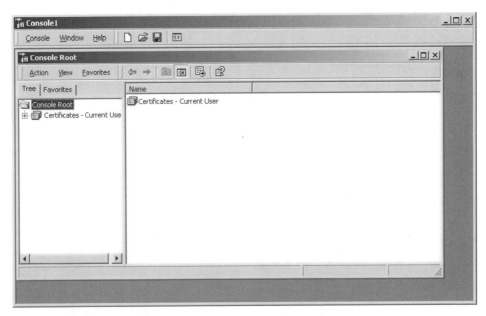

FIGURE 10.13 The console window with certificates selected for the current user.

import and browse to either ther temporary certificate you created or the one included on the CD-ROM for this book. Figure 10.14 shows the certificate registered on the author's machine.

Now that the client system has the certificate installed, you're ready to read the WSDL to make the proxy.

HTTPS and the Proxy

When executing Web Services under HTTPS, you need to ensure that your WSDL that you build your proxy against contains the proper URL. If your WSDL is generated on the fly, make sure that the WSDL changes appropriately. If you have static WSDL files, you may need to have two files: one for unsecured connections and another for secure connections.

The main piece of WSDL that changes is toward the bottom of the file and involves the address tag. You want to ensure that the WSDL generator changes appropriately when you use HTTP or HTTPS. Remember that HTTPS indicates to the Web server that SSL needs to be used for the request and response. Consider the

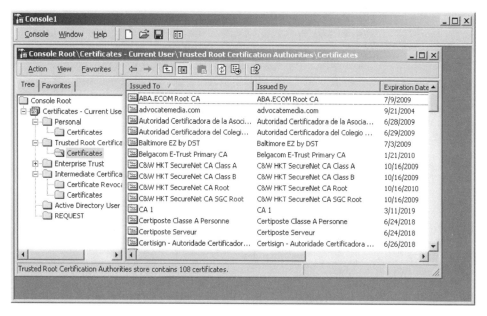

FIGURE 10.14 The console window shows the installed certificates for the author's machine.

address element for the WSDL file for one of the example Web Services that ships with Apache *Axis*.

```
<wsdl:port binding="intf:xmltoday-delayed-quotesSoapBinding"
            name="xmltoday-delayed-quotes">
  <wsdlsoap:address
     location=
     "https://HOMER/axis/servlet/AxisServlet/xmltoday-delayed-quotes" />
</wsdl:port>
```

Notice how the URL in the WSDL reflects the use of SSL with HTTPS. You need to ensure that this is present in your proxy as well.

Consider the following snippet from a generated proxy. Notice how the value for this.URL indicates SSL from the URL beginning with HTTPS. This ensures that your client accesses the Web Service through SSL.

```
/// <remarks/>
[System.Diagnostics.DebuggerStepThroughAttribute()]
```

```
[System.ComponentModel.DesignerCategoryAttribute("code")]
[System.Web.Services.WebServiceBindingAttribute
  (Name="xmltoday-  delayed-quotesSoapBinding",
   Namespace="https://HOMER/axis/servlet/AxisServlet/xmltoday" +
              "-delayed-quotes/axis/servlet/AxisS" +
              "ervlet/xmltoday-delayed-quotes")]
public class StockQuoteServiceService :
System.Web.Services.Protocols.SoapHttpClientProtocol {

/// <remarks/>
public StockQuoteServiceService() {
    this.Url =
  "https://HOMER/axis/servlet/AxisServlet/xmltoday-delayed-quotes";
}
```

Although SSL does provide encryption of requests and responses, it tends to have a very high overhead and uses a lot of memory and processor. If you need to use SSL, you may want to consider having a dedicated SSL server or use SSL as sparingly as possible.

CROSS-PLATFORM WEB SERVER SECURITY

The security that utilizes authentication does not work the same as IIS under the Apache Web server. This creates a problem for the developer who is supporting Web Service security across multiple platforms. What if you need to support authentication type security for both Java and C#? The trick is to have one service proxy the request to another behind the firewall, or to use HTTPS for the two servers to communicate directly with each other.

In this scenario, you may have a large implementation of Java Web Services up and running, but you need to support authentication under *.NET*. Because the security schemes for IIS and Apache are largely incompatible, you could still support this by setting up a server running *.NET* and IIS and then using its authentication system to protect requests to your Java Web Services by having *.NET* call these services once IIS authenticates the request. This may also be useful if the enterprise has Web Services that are Java based running somewhere behind the firewall but the Web team only supports Microsoft products. Figure 10.15 shows the *.NET* Web

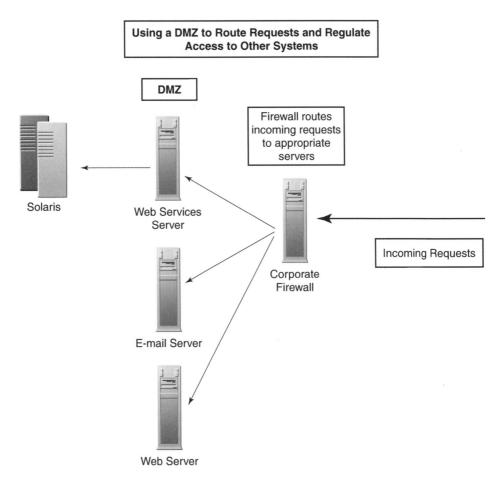

FIGURE 10.15   A *.NET* Web Services implementation proxying requests to a Java Web Services implementation behind the firewall.

Service proxying requests behind a firewall to Java Web Services running in an enterprise setting. Notice that by having a server in the DMZ, the firewall determines where requests go and what server's machines have access to. Figure 10.16 shows two Web servers operating within the *Demilitarized Zone* (DMZ) and proxying requests to each other based on the system (*.NET* or Java) calling the objects.

The following code snippet is from a *.NET* Web Service that calls a Java Web Service that comes with the Apache *Axis* distribution. With Axis' ability to create

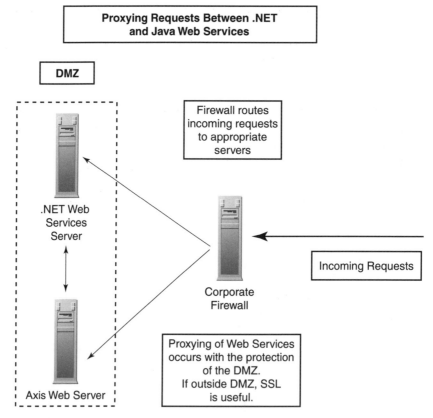

FIGURE 10.16 IIS and Apache Web servers sitting in the DMZ proxying requests to one another depending on the consumer who makes the request.

WSDL that is compatible with *.NET*, creating a C# Web Service that calls an *Axis* Web Service is very easy. Once the Web Service is written, configure IIS as shown earlier in the chapter by choosing either *Windows* Integrated Login or Basic Authentication. When you try to preview the Web Service through the browser, a dialogue box pops up and asks you for a username and password.

```
[WebMethod(Description="This method gets a stock symbol, passes it
                       to a java ws and returns the value")]
    public float getDelayedQuote(String symbol)
    {
      float myValue;
```

```
homer.ApacheDemoStockQuote myDemoStockQuote = new
homer.ApacheDemoStockQuote();
myValue = myDemoStockQuote.getQuote(symbol);
return myValue;
}
```

The previous snippet is typical of the other C# Web Services shown in the book except, instead of performing a function, it sends a request to a Java Web Service and returns the response. Note that the Web Service could call another *.NET* Web Service as well.

For proxying to work effectively, the two systems need to be somewhat close to each other or at least connected by a 100-megabit network. If any Internet tunneling occurs or a system connects through a T-1 connection, the proxying of Web Service requests will be very slow.

The following snippet is from a consumer written in C#. Notice that this event occurs in the `method` `button1_Click` indicates that this code comes from a GUI-based application.

Right before the code makes the first method code, the code creates the authentication credentials. So the consumer calls one Web Service with the appropriate credentials and the responding service then sends the request to the Java Web Service.

```
private void button1_Click(object sender, System.EventArgs e)
    {
     float returnValue;
     NetId.StockProxyService callProxy = new
     NetId.StockProxyService();
     callProxy.Credentials =
     new System.Net.NetworkCredential("Some User","password");
     returnValue = callProxy.getDelayedQuote();
     textBox1.Text = returnValue;
    }
```

This solves the security incompatibility because you use *Windows* Web Services to communicate with *Windows* consumers such as *ASP.NET*. Then through the back door you communicate with Apache Web Services via HTTPS. (Remember it is possible for *Tomcat* and *.NET* to run on the same *Windows* server.)

THE FUTURE OF *SOAP* AND XML SECURITY

As Web Services developed, people who actually needed to implement Web Services quickly noticed the lack of security present in the standard and released software. At first, many believed that security was already in place because the *SOAP* standard utilized a port that was already available through the firewall. However, if you consider the information that you move around on the Internet such as banking or credit card information, you realize security is necessary for the transmitted data in case of interception.

There is also a need for security in XML beyond *SOAP*. As XML becomes a standard way to represent data, there is a need to protect the data in the documents and ensure the documents come from the expected source. XML Encryption and XML Signature solve this problem by providing a means to accomplish these goals.

As of this writing, it seems that the Web Services community is still scrambling to create standards for software developers to follow in their implementations. The Apache Group developed an XML encryption library, but they admit that it doesn't really follow any standard. Microsoft, in the Fall of 2002, released their "Web Service Development Kit," which adheres to the current standards for XML encryption and signatures put out by the World Wide Web Consortium (W3C).

XML Encryption

XML encryption involves taking the contents of elements within the XML document and encrypting them so the contents are unusable to someone who doesn't have the decryption key. Consider the following XML document from Chapter 2.

```
<?xml version="1.0" ?>
<BOOK>
  <TITLE>Cross Platform Web Services</TITLE>
  <PAGECOUNT>4000</PAGECOUNT>
  <AUTHOR>Brian Hochgurtel</AUTHOR>
  <PUBLISHER>Charles River Media</PUBLISHER>
</BOOK>
```

There really isn't anything critical in the document, but if you add some more personal information, such as Social Security Number, that definitely needs to be encrypted. XML encryption involves adding elements that have encoding information that is generated by software running in the application that creates the XML. The encrypted information then looks like the following.

```
            <?xml version="1.0" ?>
<BOOK>
  <TITLE>Cross Platform Web Services</TITLE>
  <PAGECOUNT>4000</PAGECOUNT>
  <AUTHOR>Brian Hochgurtel</AUTHOR>
  <PUBLISHER>Charles River Media</PUBLISHER>
   <SSN Owner="Author">
    <EncryptedData
       xmlns="http://www.w3.org/2001/04/xmlenc#"
       Type="http://www.w3.org/2001/04/xmlenc#Content">
       <CipherData>
           <CipherValue>B11695C22</CipherValue>
       </CipherData>
    </EncryptedData>
   </SSN>
 </BOOK>
```

Now the part of the XML document containing the more sensitive information is encrypted, and thus protects that information from prying eyes. A *SOAP* message is an XML document and could contain encrypted information. This solves some of the security weaknesses mentioned earlier in this chapter. It also provides a means of encrypting only the data that requires protection. When using SSL, the entire transaction occurs under encryption—which uses a lot of overhead on the server. With XML encryption, only the sensitive data uses encryption.

XML Signature

XML signatures provide a means of determining who sent a particular document. This way you confirm that you know where the document comes from and you prevent a user from denying that they sent a particular document.

An XML signature involves signing a resource available on the Web with a unique key.

If you wish to sign an XML document, it would need to reside at a URL such as *http://www.advocatemedia.com/documents/test.xml*. The following snippet identifies the document to be signed.

```
<Reference
URI="http://www.advocatemedia.com/documents/test.xml">
</Reference>
```

Software then generates a unique digest based on the content of the resource and the results become part of the Reference element.

```
    <Reference
URI="http://www.advocatemedia.com/documents/test.xml">
  <Digest
    method="http://www.w3.org/2000/09/xmldsig#sha1"/>
  <DigestValue>
      ff573499e920e34acc38659bd991ce54
  </DigestValue>
</Reference>
```

To verify the signature, the key for the certificate used to sign the resource needs to be part of the document as well. This belongs within the KeyInfo element after the reference information. This example used the X509 certificate used in the SSL examples earlier in the chapter.

```
    <Reference
URI="http://www.advocatemedia.com/documents/test.xml">
  <Digest
    method="http://www.w3.org/2000/09/xmldsig#sha1"/>
  <DigestValue>
      ff573499e920e34acc38659bd991ce54
  </DigestValue>
</Reference>
    <KeyInfo>
    <X509Data>
        <X509SubjectName>
            Brian Hochgurtel
        </X509SubjectName>
        <X509Certificate>
        -----BEGIN RSA PRIVATE KEY-----
    MIICXQIBAAKBgQC14RIJZ6cdZSbEZXKMJvXlG2rgRJOSEQRUsOMzSTmrt0v9n52U
    t/adQtgayhTHlWczGUweJ3QUzn3G+QduZq414tbqXqiZtkhA3EcUSGxUSR9W
    5K70
    NR3MMRzyOsMIiOS4+e3Hdt7Ersa014TgvUPfsN4MwXvltGJmshZ+ldBV8QID
    AQAB
    AoGAbfIVlivbgNCBw91ThleS86FEVTf/QSAaTxvy7DDKtPwD6thPSPSAFwau
    Xltk
    flXZYbFcKypMaLt+mwY1MM7PZezNZdLbO22nZZaFIqDEKaJDRKiOiXBGmffJ
    3J/X
```

```
      1h5yXYZHmom6EbwKaWTjIeLKPT8G6OjGnXjd9aDaLOPlVvUCQQDYd6/+1f3
       RM/Er
      gw9UJ09lN8MbtbOc8aeP6eEaecBcr1KM2wE/HQm/AW583oye3m19GyChk2c
       SOdk/
      B4upgil7AkEA1xhOD92hjWY6FSB36VmzctKNBUYa8YsIuzQB9kZPS27HFcXJ
       6W9Y
      Rqlw36GODwQmQrNEAhVQBvyW1vrC4o6UgwJBAMEvT6omYDbsHDew52U7D+hN
       M5rv
      Pq8uG1ScbYCrV7lf3lRGv34L9D66kFhwZR8DcsNMCnsoibwCVJejrEjDGTEC
       QQCV
      yEoLyFVIhuhpb9uwtpM8oRwskP4QN7ZTzkqTebCcIb8nDT2mfa/mPPXp9MvT
       LRuL
      lRQFs1uwEdLkT2jIpWsLAkAOloRPPjMjS1RoXROUdsoDSqvTObVPjgJZl/h7
       6WW4
      kVDB4JwmMAplr2tiC8Uj1izLoxnMfZRoBE04fibc6SQu
          -----END RSA PRIVATE KEY-----
        </X509Certificate>
      </X509Data>
    </KeyInfo>
```

Now that you have the information for the resource, you need introduce the `SignedInfo` parent element and some information regarding the methods used to sign the information.

```
    <SignedInfo Id="Advocate Media">
   <CanonicalizationMethod
    Algorithm="http://www.w3.org/TR/2001/
       Rec-xml-cln4n-20010315/>
         <Digest
       method="http://www.w3.org/2000/09/xmldsig#shal"/>
    <Reference
   URI="http://www.advocatemedia.com/documents/test.xml">
     <Digest
       method="http://www.w3.org/2000/09/xmldsig#shal"/>
     <DigestValue>
        ff573499e920e34acc38659bd991ce54
     </DigestValue>
   </Reference>
       <KeyInfo>
       <X509Data>
          <X509SubjectName>
             Brian Hochgurtel
```

```
                  </X509SubjectName>
                  <X509Certificate>
                  -----BEGIN RSA PRIVATE KEY-----
      MIICXQIBAAKBgQC14RIJZ6cdZSbEZXKMJvXlG2rgRJOSEQRUsOMzSTmrt0v9n52U
        t/adQtgayhTHlWczGUweJ3QUzn3G+QduZq414tbqXqiZtkhA3EcUSGxUSR9
        W5K70
        NR3MMRzyOsMIiOS4+e3Hdt7Ersa014TgvUPfsN4MwXvltGJmshZ+ldBV8QID
        AQAB
        AoGAbfIVlivbgNCBw91ThleS86FEVTf/QSAaTxvy7DDKtPwD6thPSPSAFwauX
        ltk
        flXZYbFcKypMaLt+mwY1MM7PZezNZdLb022nZZaFIqDEKaJDRKiOiXBGmffJ
        3J/X
        1h5yXYZHmom6EbwKaWTjIeLKPT8G6OjGnXjd9aDaLOPlVvUCQQDYd6/+1f3R
        M/Er
        gw9UJ09lN8Mbtb0c8aeP6eEaecBcr1KM2wE/HQm/AW583oye3m19GyChk2c
        SOdk/
        B4upgil7AkEA1xhOD92hjWY6FSB36VmzctKNBUYa8YsIuzQB9kZPS27HFcXJ
        6W9Y
        Rqlw36GODwQmQrNEAhVQBvyW1vrC4o6UgwJBAMEvT6omYDbsHDew52U7D+hN
        M5rv
        Pq8uG1ScbYCrV7lf3lRGv34L9D66kFhwZR8DcsNMCnsoibwCVJejrEjDGTEC
        QQCV
        yEoLyFVIhuhpb9uwtpM8oRwskP4QN7ZTzkqTebCcIb8nDT2mfa/mPPXp9MvT
        LRuL
        lRQFs1uwEdLkT2jIpWsLAkAOloRPPjMjS1RoXROUdsoDSqvTObVPjgJZl/h7
        6WW4
        kVDB4JwmMAplr2tiC8Uj1izLoxnMfZRoBEO4fibc
        6SQu
                  -----END RSA PRIVATE KEY-----
              </X509Certificate>
          </X509Data>
      </KeyInfo>
  </SignedInfo>
```

Note that the CanonicalizationMethod indicates the method used to create this signature.

Utilizing this signature may involve signing a Web Service or a consumer. The provider of the Web Service is saying that the service is truly provided by his organization. The consumer states that this is really his organization accessing the service.

Signing SOAP Messages

The Web Services Development Kit (WSDK), which is an early software release from Microsoft, provides you with a means to sign a *SOAP* request within *.NET*. The methods provided allow you to prove identity or time. See Appendix B for download instructions.

To use these methods, you need to register the certificate using the "Management Console" shown earlier in the chapter. The methods used are similar to using Authentication methods used in previous examples.

The following snippet creates a token object by loading the certificate registered with the "Management Console," and if there is the certificate is added to the request of the Web Service.

```
X509SecurityToken token = GetSecurityToken();
    if (token == null) {
    throw new ApplicationException
    ("No key provided for signature.");
    }
// Add the signature element to a security section on
// the request to sign the request
requestContext.Security.Tokens.Add(token);
requestContext.Security.Elements.Add
(new Signature(token));
//Call Web Service Here
```

Now the request to the Web Service will be signed with the certificate.

In the Web Service, the code needs to be written to ensure that the certificate information is handled and valid. The following snippet demonstrates C# code examining the certificate contents. If no certificate loads, the request is returned immediately. The statement that creates the `mySignature` object looks at the *SOAP* request for signatures. If the token then looks like the one you're searching for, authorization can occur.

```
//If no security information we can reject the request.
 if ( context.Security.Tokens.Count == 0 )
            return false;

 Signature mySignature = context.Security.Elements[i] as Signature
 X509SecurityToken myCertToken =
 mySignature.SecurityToken as X509SecurityToken;
```

```
if (myCertToken = "SomeCertificateValue") {
    //perform the authorization
}
```

This is just one more way to provide security for Web Services.

CONCLUSION

The author worked on a team that created a Web Service software package that interacted with databases. The group believed that it was a great product, and it contained some extra features like the ability to transform the *SOAP* response with XSL. The first question out of every potential customer's mouth was about security, and the team hadn't provided any code or interface for securing these transactions. How did a group of very smart developers overlook this? As the team went back and examined the *SOAP* and WSDL standards closely, the realization was that the standards groups had overlooked security as well. Several development teams probably had the same experience.

The security methods described in this chapter show how much of an after-thought security was with Web Services. Using "Authorization," "Authentication," and SSL all work, but the implementations seem haphazard and confusing. As the standards groups move forward, XML signature and encryptions are steps in the right direction. The developer still must wonder if there is technology to truly secure a Web Services implementation.

11

Practical Application of Web Services

257

The code in previous chapters centers on simple examples, such as the Sim-pleStockQuote, to examine how the technology works rather than the practical application of Web Services. This chapter demonstrates a more complicated Web Service that returns different types and contains more than one method. This helps show you how to return various types of variables such as a float or string. There are some subtleties here. Additionally, this chapter allows you to review several of the subjects covered in the second half of this book such as security and creating proxies for Java code.

The examples here focus on a service that converts U.S. dollars into foreign currency, and the sections in this chapter take a four-stage approach to engineering this service, including: design, develop, test, and deploy. In the design stage, you consider the requirements for the service such as security and platforms supported. In the development stage, you write the code to meet the requirements that were originally set. The testing stage allows you to test the code you created in the development stage. This is usually the time when management adds more requirements that push you back into development for a little longer. The deployment stage involves configuring the hardware and networking necessary to implement the Web Service.

THE DESIGN STAGE

In the design stage, a development team gathers requirements from the user base. In this example, users need a currency exchange system that converts money into certain foreign denominations, converts foreign money into dollars, and returns the name of a particular currency. The example covered in this chapter does all this for money from Argentina, Australia, China, Japan, Switzerland, Mexico, and the UK.

Another consideration is to determine the platform from which to execute your Web Services. Should it be a Java-based or *.NET* Web Service? Luckily, the fact that *Axis* and *.NET* create WSDL that both utilize to create proxies for their clients gives you a lot of flexibility when making this decision. It really depends on the hardware, software, and development talent that's available to you. Java is a more mature language and seems less prone to the annoying bugs that C# occasionally has, but as Microsoft releases *Solution Packs*™ many of these bugs will disappear and *.NET* will be more stable. Support may be another consideration when choosing a technology. *.NET* has Microsoft's huge infrastructure of support teams and developers standing behind their products whereas with Apache products you need to post a message to a news group with the hopes that someone replies. Although there are plenty of supported Java-based Web Service implementations available from companies such as Cape Clear Software or BEA Systems.

The exchange rates normally need to be dynamic, such as a call to a database, but for the sake of simplifying the example, the exchange rates will be hard coded.

Once you gather the programmatic requirements, you need to consider the consumers using your services. Are the consumers applications? Or are they Web pages? Or both?

This example assumes that your user wants both Web pages and applications, and this decision directly affects how you test the Web Service because you need to test both.

Once you've gathered all these requirements, you're ready to begin development.

THE DEVELOPMENT STAGE

In this stage, the requirements are all gathered. Now it's a matter of actually implementing the code. The author chose to implement the service in Java for no other reason other than because a choice had to be made.

The Java Web Service

The following Java Web Service is meant for Apache *Axis*, which may be obvious because of the lack of import statements. All the other implementations covered in this book require import statements, but when you create a Web Service for *Axis* you just need to put a *jws* extension on the file and place it in the webapps directory. Remember, you need to make a request to it so *Axis* compiles it.

The example starts by defining the class and the first method. The method returnUSDollarEquiv takes the country name and the currency amount and converts to the United States currency. It's simply a long if-then-else statement. The returnForeignEquiv returns the equivalent to U.S. dollars, and the final method, returnCurrencyName, returns the name of the currency of the foreign country.

```
public class MoneyExchange {

    public double returnUSDollarEquiv
```

```
    (String countryName, float quantity) {

    double totalValue = 0;

    //currency values as of 10/5/02
    if (countryName.equals("Argentina")) {
        totalValue = .2663 * quantity;
    } else if (countryName.equals("Australia")) {
        totalValue = .54 * quantity;
    } else if (countryName.equals("China")) {
        totalValue = .128 * quantity;
    } else if (countryName.equals("Japan")) {
        totalValue = .0081 * quantity;
    } else if (countryName.equals("Mexico")) {
        totalValue = .0988 * quantity;
    } else if (countryName.equals("Swiss")) {
        totalValue = .675 * quantity;
    } else if (countryName.equals("UK")) {
        totalValue = 1.568 * quantity;
    } else {
        totalValue = -1;
    }

return totalValue;
}

 public double returnForeignEquiv
 (String countryName, float quantity) {

   double totalValue = 0;

   //currency values as of 10/5/02
   if (countryName.equals("Argentina")) {
      totalValue = 3.755 * quantity;
   } else if (countryName.equals("Australia")) {
      totalValue = 1.83 * quantity;
   } else if (countryName.equals("China")) {
      totalValue = 8.29 * quantity;
   } else if (countryName.equals("Japan")) {
      totalValue = .0081 * quantity;
   } else if (countryName.equals("Mexico")) {
      totalValue = 10.12 * quantity;
   } else if (countryName.equals("Swiss")) {
```

```
        totalValue = 1.4782 * quantity;
    } else if (countryName.equals("UK")) {
        totalValue = .63784 * quantity;
    } else {
        totalValue = -1;
    }

    return totalValue;
}

public String returnCurrencyName(String countryName)
{
    String name = null;

    if (countryName.equals("Argentina")) {
        name = "Peso";
    } else if (countryName.equals("Australia")) {
        name = "Dollar";
    } else if (countryName.equals("China")) {
        name = "Renminbi";
    } else if (countryName.equals("Japan")) {
        name = "Yen";
    } else if (countryName.equals("Mexico")) {
        name = "Peso";
    } else if (countryName.equals("Swiss")) {
        name = "Franc";
    } else if (countryName.equals("UK")) {
        name = "Pound";
    } else {
        name = "No match";
    }

    return name;

}
}
```

This is a good example because you need to pass-in values that are floats and strings to the service. You will find that this is a little tricky when you create Web page consumers.

Java Test Client

As discussed in Chapter 8, creating a proxy for a Java Web Service makes the client code cleaner and simpler, and allows you to develop a client quickly. You can use a tool such as *WebServiceStudio* to test your server, but this tests the service only from the *.NET* environment. So it is necessary to write a quick client test just to ensure you can connect from Java.

The first step is to generate the proxy with the following commands.

```
java org.apache.axis.wsdl.WSDL2Java
http://localhost:8080/axis/MoneyExchange.jws?wsdl
```

Remember that to use this command, the environment on your PC needs to be configured in such a way that all the libraries in *Axis* are available in your environment.

When this command executes it generates the following files:

```
MoneyExchange.java
MoneyExchangeSoapBindingStub.java
MoneyExchangeService.java
MoneyExchangeServiceLocator.java
```

`MoneyExchange.java` shows the methods available from the Money Exchange Web Service. It is simply an interface definition that defines the methods that other classes in this proxy need to utilize such as `MoneyExchangeService`. Notice that there isn't any code here that executes. Remember to compile all these classes before trying to use them in client code.

```
/**
 * MoneyExchange.java
 *
 * This file was auto-generated from WSDL
 * by the Apache Axis WSDL2Java emitter.
 */

package localhost;

public interface MoneyExchange extends java.rmi.Remote
{
    public java.lang.StringreturnCurrencyName
```

```
        (java.lang.String countryName) throws
        java.rmi.RemoteException;

        public double returnForeignEquiv
        (java.lang.String countryName, float quantity)
         throws java.rmi.RemoteException;

        public double returnUSDollarEquiv
        (java.lang.String countryName, float quantity)
        throws java.rmi.RemoteException;
    }
```

MoneyExchangeService reveals the methods for calling the service encapsulated by the proxy created here.

```
    /**
     * MoneyExchangeService.java
     *
     * This file was auto-generated from WSDL
     * by the Apache Axis WSDL2Java emitter.
     */

    package localhost;

    public interface MoneyExchangeService extends
    javax.xml.rpc.Service {
       public String getMoneyExchangeAddress();

        public localhost.MoneyExchange getMoneyExchange()
        throws javax.xml.rpc.ServiceException;

        public localhost.MoneyExchange
        getMoneyExchange(java.net.URL portAddress)
        throws javax.xml.rpc.ServiceException;
    }
```

MoneyExchangeServiceLocator stores the location of the Web Service and then handles any errors if the service can't be reached. But unlike the comments the generator put in the code state, this is unlikely because you used WSDL that resides at the location of the Web Service. If you can reach that, you can probably reach the service.

```java
/**
 * MoneyExchangeServiceLocator.java
 *
 * This file was auto-generated from WSDL
 * by the Apache Axis WSDL2Java emitter.
 */

package localhost;

public class MoneyExchangeServiceLocator extends
org.apache.axis.client.Service
implements localhost.MoneyExchangeService {

// Use to get a proxy class for MoneyExchange
private final java.lang.String MoneyExchange_address =
"http://localhost:8080/axis/MoneyExchange.jws";

 public String getMoneyExchangeAddress() {
    return MoneyExchange_address;
 }

 public localhost.MoneyExchange getMoneyExchange()
 throws javax.xml.rpc.ServiceException {
    java.net.URL endpoint;
    try {
       endpoint = new
       java.net.URL(MoneyExchange_address);
    }
    catch (java.net.MalformedURLException e) {
       return null; // unlikely as URL was validated
                    // in WSDL2Java
    }
    return getMoneyExchange(endpoint);
 }

 public localhost.MoneyExchange getMoneyExchange
 (java.net.URL portAddress) throws
  javax.xml.rpc.ServiceException {
   try {
       return new
       localhost.MoneyExchangeSoapBindingStub
       (portAddress, this);
   }
```

```
      catch (org.apache.axis.AxisFault e) {
          return null; // ???
      }
   }

   /**
    * For the given interface, get the stub
    *implementation.
    * If this service has no port for the given
    *interface,
    * then ServiceException is thrown.
    */

   public java.rmi.Remote getPort
   (Class serviceEndpointInterface) throws
    javax.xml.rpc.ServiceException {
    try {
        if
        (localhost.MoneyExchange.class.isAssignableFrom
        (serviceEndpointInterface)) {
            return new
            localhost.MoneyExchangeSoapBindingStub
            (new java.net.URL(MoneyExchange_address),
            this);
        }
    }
    catch (Throwable t) {
        throw new javax.xml.rpc.ServiceException(t);
    }
    throw new javax.xml.rpc.ServiceException
    ("There is no stub implementation for the interface"
    +(serviceEndpointInterface == null ? "null" :
      serviceEndpointInterface.getName()));
   }
 }
```

MoneyExchangeSoapBindingStub.java contains the actual client code. Look toward the bottom and you'll find client code just like that was written in Chapters 7 and 8. Except this time the proxy generator created it for you. If you look at the proxy code carefully, you'll notice that it is much like the WSDL XML in some ways. It starts out by defining each of the details of the service and works its way down to the actual call.

```java
/**
 * MoneyExchangeSoapBindingStub.java
 *
 * This file was auto-generated from WSDL
 * by the Apache Axis WSDL2Java emitter.
 */

package localhost;

public class MoneyExchangeSoapBindingStub extends
org.apache.axis.client.Stub implements
localhost.MoneyExchange {

private java.util.Vector cachedSerClasses = new
java.util.Vector();
private java.util.Vector cachedSerQNames = new
java.util.Vector();
private java.util.Vector cachedSerFactories = new
java.util.Vector();
private java.util.Vector cachedDeserFactories = new
java.util.Vector();

public MoneyExchangeSoapBindingStub() throws
org.apache.axis.AxisFault {
    this(null);
}

public MoneyExchangeSoapBindingStub
(java.net.URL endpointURL, javax.xml.rpc.Service
service)
throws org.apache.axis.AxisFault {
   this(service);
   super.cachedEndpoint = endpointURL;
}

public
MoneyExchangeSoapBindingStub
(javax.xml.rpc.Service service) throws
 org.apache.axis.AxisFault {
    try {
        if (service == null) {
            super.service = new
            org.apache.axis.client.Service();
```

```
        } else {
            super.service = service;
        }
    }
    catch(java.lang.Exception t) {
        throw org.apache.axis.AxisFault.makeFault(t);
    }
}

private org.apache.axis.client.Call createCall() throws
 java.rmi.RemoteException {
    try {
        org.apache.axis.client.Call call =
        (org.apache.axis.client.Call)
        super.service.createCall();
        if (super.maintainSessionSet) {

        call.setMaintainSession(super.maintainSession);
        }
        if (super.cachedUsername != null) {
           call.setUsername(super.cachedUsername);
        }
        if (super.cachedPassword != null) {
           call.setPassword(super.cachedPassword);
        }
        if (super.cachedEndpoint != null) {
           call.setTargetEndpointAddress
          (super.cachedEndpoint);
        }
        if (super.cachedTimeout != null) {
           call.setTimeout(super.cachedTimeout);
        }
        java.util.Enumeration keys =
        super.cachedProperties.keys();
        while (keys.hasMoreElements()) {
            String key = (String)
                keys.nextElement();
            if(call.isPropertySupported(key))
                call.setProperty(key,
                super.cachedProperties.get(key));
            else
                call.setScopedProperty(key,
                super.cachedProperties.get(key));
```

```
            }
            // All the type mapping information is
            // registered
            // when the first call is made.
            // The type mapping information is

            // actually registered in
            // the TypeMappingRegistry of the service,
            // which
            // is the reason why registration is only
            // needed for the first call.
            synchronized (this) {
                if (firstCall()) {
                 // must set encoding style before
                  registering serializers
                  call.setEncodingStyle
                 (org.apache.axis.Constants.URI_SOAP11_ENC);
                    for (int i = 0; i <
                        cachedSerFactories.size(); ++i) {
                        Class cls = (Class)
                        cachedSerClasses.get(i);
                        javax.xml.namespace.QName qName =
                        (javax.xml.namespace.QName)
                         cachedSerQNames.get(i);
                         Class sf = (Class)
                           cachedSerFactories.get(i);
                         Class df = (Class)

                         cachedDeserFactories.get(i);
                         call.registerTypeMapping
                         (cls, qName, sf, df, false);
                    }
                }
            }
            return call;
        }
        catch (Throwable t) {
            throw new org.apache.axis.AxisFault
            ("Failure trying to get the Call object", t);
        }
    }

    public java.lang.String
```

```
returnCurrencyName(java.lang.String countryName)
throws java.rmi.RemoteException{
if (super.cachedEndpoint == null) {
    throw new
    org.apache.axis.NoEndPointException();
}
org.apache.axis.client.Call call = createCall();
javax.xml.namespace.QName pOQName = new
javax.xml.namespace.QName("", "countryName");
call.addParameter
(pOQName, new javax.xml.namespace.Qname
("http://www.w3.org/2001/XMLSchema", "string"),
java.lang.String.class,
javax.xml.rpc.ParameterMode.IN);
call.setReturnType(new javax.xml.namespace.Qname
("http://www.w3.org/2001/XMLSchema", "string"));
call.setUseSOAPAction(true);
call.setSOAPActionURI("");
call.setOperationStyle("rpc");
call.setOperationName
(new javax.xml.namespace.Qname
("http://localhost:8080/axis/MoneyExchange.jws",
 "returnCurrencyName"));

Object resp = call.invoke
(new Object[] {countryName});

if (resp instanceof java.rmi.RemoteException) {
    throw (java.rmi.RemoteException)resp;
}
else {
    try {
        return (java.lang.String) resp;
    } catch (java.lang.Exception e) {
      return (java.lang.String)
      org.apache.axis.utils.JavaUtils.convert
      (resp, java.lang.String.class);
    }
  }
}
        /* And much more generated code */
```

The simple Java client imports the proxy code, and then creates a location object which stores information about where the Web Service resides. Then creating a MoneyExchange object involves calling the getMoneyExchange method from the location object.

```java
import localhost.*;

public class TestMoneyExchange {

    public static void main(String [] args)
    throws Exception {

    //init value passed back from service
    double returnedValue = 0;
    String currencyName = null;

    MoneyExchangeServiceLocator myMoneyExchangeLocation =
    new MoneyExchangeServiceLocator();

    localhost.MoneyExchange myMoneyExchange =
    myMoneyExchangeLocation.getMoneyExchange();

    //get currency name
    currencyName = myMoneyExchange.returnCurrencyName
    ("Japan");
    System.out.println("The name of the currency is:"
    + currencyName);

    //get equiv
    returnedValue = myMoneyExchange.returnForeignEquiv
    ("Japan",1000);
     System.out.println("The Japanese Equivalent is:"
     + returnedValue);

     //get other equiv
     returnedValue =
     myMoneyExchange.returnUSDollarEquiv("Japan",1000);
     System.out.println("The American Equivalent is:"
     + returnedValue);
    }
   }
```

Remember this test client just ensures that you can test the Web Service from the desired platform. There needs to be more extensive testing of the Web Service, and this gets covered in greater detail later in the chapter.

JSP Consumer

JSP provides an easier way than a servlet to integrate Java code into a Web page because code mixes with the HTML elements. This way a developer does not need to be involved to change something simple like a title. Java code can be abstracted into a JSP custom tag or simply coded in a Scriptlet where code appears between <% %> elements in a page.

In this first JSP example, the page starts out by importing the classes from the package localhost which were created by the *WSDL2Java tool*. A scriptlet then instantiates the location and the myMoneyExchange objects so the call to the returnCurrencyName method can occur. The name of the currency is then displayed with <%= %> tags so the value of the currencyName variable actually displays in the web page.

Remember that to use the code generated by the WSDL2Java tool, *you need to place the Java class files in the* web apps WEB-INF/classes *directory. This is where* Tomcat *looks for classes to load when it executes.*

```
<HTML>
<HEAD>
    <TITLE>This is a simple JSP Client</TITLE>
</HEAD>
<BODY>
<!-- import the proxy classes -->
<%@ page import="localhost.*" %>

<%
//init value passed back from service
double returnedValue = 0;
String currencyName = null;

MoneyExchangeServiceLocator myMoneyExchangeLocation =
new MoneyExchangeServiceLocator();

localhost.MoneyExchange myMoneyExchange =
myMoneyExchangeLocation.getMoneyExchange();
```

```
        //get currency name
        currencyName =
        myMoneyExchange.returnCurrencyName("Japan");


    %>
    <h2>
        The currency name is <%= currencyName %>
    </h2>
    </BODY>
    </HTML>
```

The previous example is a simple JSP page that allows you to test the ability to contact the Web Service, but really would be useful to any of your clients who sent you requirements. It doesn't allow you to choose a particular country to convert to and you can't specify an amount. The next JSP example takes the functionality one step further by being more interactive.

Again, the first chunk of code imports the libraries you need. Notice that in this example there is an import for `java.text.*`. This provides the JSP page the ability to format the output of the some of the method calls.

Following the import statement, the code creates several variables so that the form can be processed differently depending on whether the page first loads or receives a request through a `Post` method. Notice that `dollarsConverter` receives the value of the `getParameter` method of the `request` object. The first time the page loads this won't have a value, so the code that appears right after the variable definitions will not execute.

The next chunk of code again looks to see if data has been submitted to the form, and if it has it reads the `countryName` value from the request and coverts the `dollarsConverter` variable to the float variable dollars using methods from the `Float` class.

An if statement then checks to ensure that both variables that need to be submitted by the form, `countryName` and `dollars`, have value before executing the code after the conditional. This prevents errors that occur when the methods to the service get called before the variables they pass get populated. If the variables do indeed have value, the calls are made to the Web Service just like in the simple Java client shown earlier, and the values are then displayed in the HTML that follows.

Then there is a `form` tag whose action sends the information back to this JSP page for processing and display. The remaining HTML elements set up a text box and a select HTML tags so the user can select country and enter the quantity of currency they wish to convert.

```
<HTML>
<HEAD>
    <TITLE>Convert to Other Currency</TITLE>
</HEAD>
<BODY>
<%@ page import="localhost.*, java.text.*" %>

<!-- The following handles the response -->
<!-- Get parameters passed from get or post -->
<%
String dollarsConverter = null;
dollarsConverter = request.getParameter("dollars");
String countryName = null;
String currencyName = null;
float dollars = 0;


if(dollarsConverter != null)  {
    countryName = request.getParameter("countryName");
    dollars =
    Float.valueOf(dollarsConverter).floatValue();
}

%>

<!-- Call methods with values passed in from form -->
<%

 if ((countryName != null) && (dollars > 0)) {
     double foreignValue = 0;

 MoneyExchangeServiceLocator myMoneyExchangeLocation
 = new MoneyExchangeServiceLocator();

 localhost.MoneyExchange myMoneyExchange
 = myMoneyExchangeLocation.getMoneyExchange();

 //get currency name
 currencyName =
 myMoneyExchange.returnCurrencyName(countryName);

 foreignValue =
   myMoneyExchange.returnForeignEquiv
   (countryName,dollars);
```

```
String formattedForeignValue =
  NumberFormat.getCurrencyInstance()
  format(foreignValue);
%>

<h3>
    The equivalent to  <%= dollars %> US Dollars
    in  <%= currencyName %> is  
    <%= formattedForeignValue %>.
</h3>
<% } %>

<!--notice that form sends data to itself -->
<form method="get" action="test2.jsp">
<h2>
    Please enter the amount of dollars and the
    country's currency you wish to convert to.
</h2>

 Dollar amount:
 <input type="text" name="dollars"><br>

 Select a country:
 <select name = "countryName">
    <option></option>
    <option>Argentina</option>
    <option>Australia</option>
    <option>China</option>
    <option>Mexico</option>
    <option>Swiss</option>
    <option>UK</option>
 </select>
 <p>

 <input type="submit"><br>

</form>

</BODY>
</HTML>
```

This example provides more of a real-world example of applying a Web Service. This page actually takes and processes information from the user that is then sent to the Web Service. Figure 11.1 shows how this form appears in *Internet Explorer*.

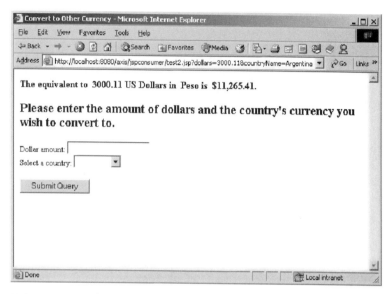

FIGURE 11.1 The JSP consumer with the drop-down list.

This is the type of form your users are likely to demand for an interface to a Web Service.

C# Test Client

Originally the Web Service in this chapter was only going to support JSP pages, but in the ever-changing world of Information Technology, "management" decided that the monetary conversion Web Service needs to be compatible with *.NET*. Even though Apache *Axis* possesses tools that generate WSDL that is reliably compatible with *.NET*, you still need to test each service from your desired language within *.NET*. This ensures that your services are truly cross platform compatible during your development stage, prevents you from discovering any glitches during deployment, and ensures that some client code actually works.

Just like with the Java code, *Visual Studio.NET* or the *.NET Framework SDK* creates a proxy for each Web Service contacted. The following C# code is the proxy for the MoneyExchange Java Web Service. If you examine the code closely, you'll find that it does many of the same things that the Java proxy code does. It contains information about the Web Service location and the types of values that need to be passed back and forth, and also generates the appropriate namespace for the client to be able to call the service. The proxy Microsoft generates, however, is much more succinct than the Java proxy.

It is not coincidence that the proxy code and the WSDL that both .NET and Axis Web Services generate are similar. Microsoft and IBM (remember that IBM started the Axis project as the "Web Service Toolkit") work very closely together on not only creating Web Service software but also toward creating standards they find useful. Remember that SOAP was originally an intellectual product of Microsoft until it was released to the standards committees.

```
//-------------------------------------------------
// <autogenerated>
//     This code was generated by a tool.
//     Runtime Version: 1.0.3705.0
//
//     Changes to this file may cause incorrect
//     behavior and will be lost if
//     the code is regenerated.
// </autogenerated>
//-------------------------------------------------
//
// Assembly WebServiceStudio Version = 1.0.3705.0
//
using System.Diagnostics;
using System.Xml.Serialization;
using System;
using System.Web.Services.Protocols;
using System.ComponentModel;
using System.Web.Services;


// <remarks/>
[System.Diagnostics.DebuggerStepThroughAttribute()]
[System.ComponentModel.DesignerCategoryAttribute
("code")]
[System.Web.Services.WebServiceBindingAttribute(
    Name="MoneyExchangeSoapBinding",
    Namespace=
```

```
    "http://localhost:8080/axis/MoneyExchange.jws")]

public class MoneyExchangeService :
System.Web.Services.Protocols.SoapHttpClientProtocol {

 /// <remarks/>
 public MoneyExchangeService() {
     this.Url =
     "http://localhost:8080/axis/MoneyExchange.jws";
 }

 /// <remarks/>

 [System.Web.Services.Protocols.SoapRpcMethodAttribute
 ("",
   RequestNamespace=
   "http://localhost:8080/axis/MoneyExchange.jws",
   ResponseNamespace=
   "http://localhost:8080/axis/MoneyExchange.jws")]
 [return:
 System.Xml.Serialization.SoapElementAttribute
 ("return")]
 public string returnCurrencyName(string countryName) {
     object[] results =
     this.Invoke("returnCurrencyName",
                 new object[] {
                 countryName});
     return ((string)(results[0]));
 }

 /// <remarks/>
 public System.IAsyncResult
 BeginreturnCurrencyName(string countryName,
 System.AsyncCallback callback, object asyncState) {
     return this.BeginInvoke
     ("returnCurrencyName", new object[] {
       countryName}, callback, asyncState);
 }

 /// <remarks/>
 public string EndreturnCurrencyName
 (System.IAsyncResult asyncResult) {
     object[] results = this.EndInvoke(asyncResult);
```

```
        return ((string)(results[0]));
    }

    /// <remarks/>
    [System.Web.Services.Protocols.SoapRpcMethodAttribute
    ("", RequestNamespace=
    "http://localhost:8080/axis/MoneyExchange.jws",
    ResponseNamespace=
    "http://localhost:8080/axis/MoneyExchange.jws")]
    [return: System.Xml.Serialization.SoapElementAttribute
    ("return")]
     public System.Double returnForeignEquiv
     (string countryName, System.Single quantity) {
        object[] results =
        this.Invoke("returnForeignEquiv", new object[] {
                    countryName, quantity});
        return ((System.Double)(results[0]));
      }

    /// <remarks/>
     public System.IAsyncResult
    BeginreturnForeignEquiv(string countryName,
                            System.Single quantity,
                            System.AsyncCallback callback,
                            object asyncState) {
        return this.BeginInvoke
       ("returnForeignEquiv", new object[] {
           countryName,
             quantity}, callback, asyncState);
    }

    /// <remarks/>
    public System.Double
    EndreturnForeignEquiv(System.IAsyncResult asyncResult)
     {
        object[] results = this.EndInvoke(asyncResult);
        return ((System.Double)(results[0]));
    }

    /// <remarks/>

    [System.Web.Services.Protocols.SoapRpcMethodAttribute
    ("",
```

```
RequestNamespace=
"http://localhost:8080/axis/MoneyExchange.jws",
ResponseNamespace=
"http://localhost:8080/axis/MoneyExchange.jws")]
[return:
System.Xml.Serialization.SoapElementAttribute
("return")]
public System.Double returnUSDollarEquiv
(string countryName, System.Single quantity) {
    object[] results = t
    this.Invoke("returnUSDollarEquiv", new object[] {
                countryName,quantity});
    return ((System.Double)(results[0]));
}

/// <remarks/>
public System.IAsyncResult
BeginreturnUSDollarEquiv
(string countryName,
 System.Single quantity,
 System.AsyncCallback callback, object asyncState) {
    return
    this.BeginInvoke("returnUSDollarEquiv",
    new object[] {
    countryName,quantity}, callback, asyncState);
}

/// <remarks/>
public System.Double
EndreturnUSDollarEquiv(System.IAsyncResult asyncResult)
{
    object[] results = this.EndInvoke(asyncResult);
    return ((System.Double)(results[0]));
}
}
```

Now that the proxy is part of your *Visual Studio.NET* project, you can create a simple test client that ensures you can call the methods from the Java Web Service. Note that in the following C# example that you are not required to create a location object first and then call the service like you did with Java. Microsoft hides that detail from you.

```
using System;

namespace TestJavaMoneyService
{

class Class1
{
[STAThread]
static void Main(string[] args)
{
String currencyName = null;
double returnedValue = 0;

 localhost.MoneyExchangeService myMoneyExchange =
new localhost.MoneyExchangeService();
currencyName =
myMoneyExchange.returnCurrencyName("Argentina");
Console.WriteLine
("The name of the money in Argentina:"
 + currencyName);

returnedValue =
myMoneyExchange.returnForeignEquiv("Argentina",12322);
Console.WriteLine
("The Argentinian equivalent is: " + returnedValue);

returnedValue =
myMoneyExchange.returnUSDollarEquiv("Argentina",433);
Console.WriteLine
("The Argentinian equivalent is: " + returnedValue);
 }
}
}
```

This C# Client tests each of the three methods available through the Web Service. It is not an extensive test but does provide the knowledge that the Web Service is available.

ASP.NET Consumer

The following *ASP.NET* example sets up a simple Web page that makes a call to the Web Service with a value that's hard-coded in the form. As with all the other examples, this example checks to see if you can actually call the Web Service within the *ASP.NET* page. Also, this example assumes that you have already added a Web reference in *Visual Studio.NET* in order to create the proxy.

The import statement in this example makes the proxy code to the Web Service available. The page definition describes which *.NET* language this page uses and information describing where the code resides and any inheritance path.

Then, in between the opening and closing script tag, lies a method that executes when a button has the method Submit_Click associated with it. Remember that you need the runat="server" attribute of the script tag in order for the code to execute. Otherwise the code gets sent to the browser as client code, and the browser just ignores because it doesn't know what to do with the information. The code then creates a myExchange object which represents the MoneyExchange Web Service and calls the returnCurrencyName method. The string variable currencyName contains the return value and is displayed by sending the value to the Text attribute of the asp:label message.

```
<%@ Import
    Namespace="MoneyConversionClient.localhost"%>
<%@ Page language="c#"
        Codebehind="WebForm1.aspx.cs"
        AutoEventWireup="false"
        Inherits="MoneyConversionClient.WebForm1" %>

<html>
  <head>
      <title>Money Exchange Client</title>
  </head>
  <body MS_POSITIONING="GridLayout">
      <script runat="server">
          protected void Submit_Click(
          Object Src, EventArgs E){
              MoneyExchangeService myExchange =
              new MoneyExchangeService();
              String currencyName =
              myExchange.returnCurrencyName("Swiss");
              Message.Text = currencyName;
          }
      </script>


      <form id="Form1" method="post" runat="server">
```

```
        <p>

        The currency name is: 
        <asp:Label ID="Message" Runat="server" /><br>
        </p>
        <input type="submit"
         value="get money name"
         onserverclick="Submit_Click"
         runat="server"
         ID="Submit1"
         NAME="Submit1"/>
        </form>
    </body>
</html>
```

Just as with the JSP examples, the first *ASP.NET* example ensures that you can actually call the service whereas this one will actually allow someone to enter values and receive useful information about exchange rates back.

In the next *ASP.NET* code example, the attributes of the import and page tags remain the same. The method within the script tags, Submit_Click, changes in this case. Instead of just containing a hard-coded value, the code actually looks at a text box and an asp:ListBox to get information to pass to the methods. Once the methods are called, the return values are displayed within the HTML.

```
<%@ Import
 Namespace="MoneyConversionClient.localhost" %>
<%@ Page language="c#"
        Codebehind="WebForm1.aspx.cs"
        AutoEventWireup="false"
        Inherits="MoneyConversionClient.WebForm1" %>
<html>
    <head>
        <title>Money Exchange Client</title>
    </head>
 <body MS_POSITIONING="GridLayout">
     <script runat="server">
        protected void Submit_Click
        (Object Src, EventArgs E){
           String currencyName = null;
           String foreignValue = null;

           String myDollars = dollars.Text;
```

```
        String myCountryName =
        countryName.SelectedItem.Text;

        float myFloatDollars =
        float.Parse(myDollars);
        MoneyExchangeService myExchange
        = new MoneyExchangeService();

        currencyName =myExchange.returnCurrencyName
        (myCountryName);
        foreignValue =
        myExchange.returnForeignEquiv
        (myCountryName,myFloatDollars).ToString();

        currencyNameLabel.Text = currencyName;
        dollars.Text = myFloatDollars.ToString();
        foreignValueLabel.Text = foreignValue;
      }
  </script>


  <form id="Form1" method="post" runat="server">
  <p>
  The equivalent to  
 <asp:Label ID="dollars" Runat="server" />
  US Dollars in  
 <asp:Label ID="currencyNameLabel"
           Runat="server" />
  is  
  <asp:Label ID="foreignValueLabel"
            Runat="server" />.

  </p>
  <h2>
     Please enter the amount of dollars and the
     country's currency you wish to convert to.
  </h2>
Dollar amount:
<input type="text" name="dollars"><br>
   Select a country:
<asp:ListBox id = "countryName" Runat=server>
  <asp:ListItem></asp:ListItem>
```

```
          <asp:ListItem>Argentina</asp:ListItem>
          <asp:ListItem>Australia</asp:ListItem>
          <asp:ListItem>China</asp:ListItem>
          <asp:ListItem>Mexico</asp:ListItem>
          <asp:ListItem>Swiss</asp:ListItem>
          <asp:ListItem>UK</asp:ListItem>
      </asp:ListBox>
      <p>
          <input type="submit"
             value="get money name"
             onserverclick="Submit_Click"
             runat="server"
             ID="Submit1"
             NAME="Submit1"/>
          </form>
        </body>
      </html>
```

Figure 11.2 contains the output of this form once some values have been input and the "Submit" button has been clicked.

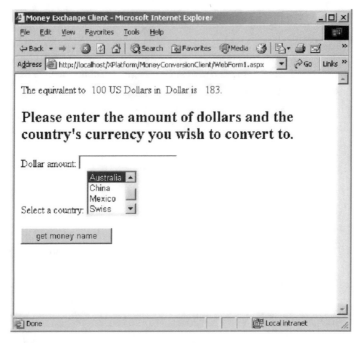

FIGURE 11.2   The *ASP.NET* client form.

This ASP.NET note example uses the "Parse" method of the float object. This takes a string read-in from the form and converts it into a float value. If you experience problems with this method or you can't compile the page due to mysterious errors on this line, you may need to download VS.NET Service Pack 2.

THE TESTING STAGE

Even though we created simple test programs during the development stage, these programs really didn't exercise the Web Services enough to test their availability and whether the methods behaved properly. Another aspect to consider during testing is the ability of your server to handle the load from the demand put on your Web Services. We begin by examining requests and responses.

Viewing Requests and Responses

With all the technology available, it's easy to forget that XML is underneath. Here's a request and response from the Money Exchange Web Service to jar your memory. Remember that the request and response can be viewed in *WebServicesStudio* and through the tools Apache provides in both *SOAP* and *Axis*. Luckily all the tools mentioned in this book work fairly well together because of the compatible WSDL the tools generate. If, however, you ever had problems communicating with a particular service, examining the request and response is a great way to look for inconsistencies. As Web Services mature, you'll find that more and more software will become compatible and the problems with making different requests and response will disappear.

The following *SOAP* request is similar to many requests viewed thus far in the book except that this is the first service that sends a float to the method. You'll notice that the quantity tag contains a type definition for xsd:float.

```
<?xml version="1.0" encoding="utf-8"?>
    <soap:Envelope
    xmlns:soap=
    "http://schemas.xmlsoap.org/soap/envelope/"
    xmlns:soapenc=
    "http://schemas.xmlsoap.org/soap/encoding/"
    xmlns:tns=
```

```
    "http://localhost:8080/axis/MoneyExchange.jws"
    xmlns:types=
    "http://localhost:8080/
     axis/MoneyExchange.jws/encodedTypes"
    xmlns:xsi=
    "http://www.w3.org/2001/XMLSchema-instance"
    xmlns:xsd=
    "http://www.w3.org/2001/XMLSchema">

    <soap:Body soap:encodingStyle=
    "http://schemas.xmlsoap.org/soap/encoding/">
   <tns:returnUSDollarEquiv>
      <countryName xsi:type="xsd:string">
      Argentina
     </countryName>
    <quantity xsi:type="xsd:float">350</quantity>
   </tns:returnUSDollarEquiv>
  </soap:Body>
 </soap:Envelope>
```

The response then returns the equivalent in US dollars as a double to the client.

```
  <?xml version="1.0" encoding="utf-8"?>
   <soapenv:Envelope
   xmlns:soapenv=
   "http://schemas.xmlsoap.org/soap/envelope/"
   xmlns:xsd=
   "http://www.w3.org/2001/XMLSchema"
   xmlns:xsi=
   "http://www.w3.org/2001/XMLSchema-instance">
   <soapenv:Body>
   <ns1:returnUSDollarEquivResponse
    soapenv:encodingStyle=
   "http://schemas.xmlsoap.org/soap/encoding/"
   xmlns:ns1=
   "http://localhost:8080/axis/MoneyExchange.jws">
   <returnUSDollarEquivReturn xsi:type="xsd:double">
   93.205
   </returnUSDollarEquivReturn>
  </ns1:returnUSDollarEquivResponse>
 </soapenv:Body>
</soapenv:Envelope>
```

Testing the Methods in General

WebServicesStudio allows you to make requests and responses to the server without having to actually create a client, as mentioned in Chapter 9. Figure 11.3 shows *WebServiceStudio* calling the methods of the Money Exchange Web Service. Using this tool allows you to test the methods and view requests and responses without having to write a program. For complete testing, however, you'll want to have a command line program or some Web pages that can test the Web Service in an automated fashion and gather statistics on which methods might be failing.

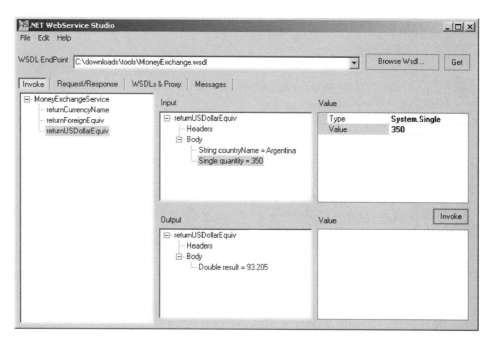

FIGURE 11.3   Using *WebServicesStudio* to test one of the Money Exchange methods.

This is a very handy tool, as long as the Web Service generates WSDL.

Creating a Test Program

Testing is something you can never do enough of, especially in a cross-platform environment. The first thing you need to test is your ability to communicate with the Web Service through the use of simple test programs as shown in previous sections of this chapter. Once that phase is complete, you need to develop a test strategy that allows you to test all the various platforms and consumer types you make available to your customer base.

Consider the following Java code. The code tests each method to see whether it returns the correct value for the attributes sent to the method. If the correct value returns, the test program prints out a message saying that the test PASSED. It also prints out messages when tests fail. This makes it easy to search a log file looking for failures. The code in this client is simply the client code used in the previous examples but here it checks each conversion to see if the proper value is returned.

```java
Import localhost.*;

public class FullTestMoneyExchange {

public static void main(String [] args) throws
Exception {

  //init value passed back from service
  double returnedValue = 0;
  String currencyName = null;

  MoneyExchangeServiceLocator myMoneyExchangeLocation
  = new MoneyExchangeServiceLocator();

  localhost.MoneyExchange myMoneyExchange =
  myMoneyExchangeLocation.getMoneyExchange();

  //get currency name for all countries
  currencyName = myMoneyExchange.returnCurrencyName
("Argentina");
  if (currencyName.equals("Peso")) {
    System.out.println("Argentina Money Name Test
    PASSED "  + currencyName);
  } else {
    System.out.println
    ("Argentina Money Name Test Failed " +
    currencyName);
  }

  currencyName =
  myMoneyExchange.returnCurrencyName("Australia");
  if (currencyName.equals("Dollar")) {
```

```
    System.out.println("Australia Money Name Test
    PASSED " + currencyName);
} else {
    System.out.println("Australia Money Name Test
    Failed " + currencyName);
}

currencyName =
myMoneyExchange.returnCurrencyName("China");
if (currencyName.equals("Renminbi")) {
    System.out.println("China Money Name Test PASSED "
    + currencyName);
} else {
    System.out.println("China Money Name Test
    Failed "  + currencyName);
}

currencyName =
myMoneyExchange.returnCurrencyName("Japan");
if (currencyName.equals("Yen")) {
    System.out.println("Japan Money Name Test PASSED "
    + currencyName);
} else {
    System.out.println("Japan Money Name Test
    Failed "  + currencyName);
}

currencyName =
myMoneyExchange.returnCurrencyName("Mexico");
if (currencyName.equals("Peso")) {
    System.out.println
    ("Mexico Money Name Test PASSED "
     + currencyName);
} else {
    System.out.println("Mexico Money Name Test Failed"
    + currencyName);
}

currencyName =
myMoneyExchange.returnCurrencyName("Swiss");
if (currencyName.equals("Franc")) {
    System.out.println("Swiss Money Name Test PASSED "
    + currencyName);
```

```java
} else {
  System.out.println("Swiss Money Name Test Failed "
  + currencyName);
}

currencyName =
myMoneyExchange.returnCurrencyName("UK");
if (currencyName.equals("Pound")) {
  System.out.println("Swiss Money Name Test PASSED "
  + currencyName);
} else {
  System.out.println("Swiss Money Name Test Failed "
  + currencyName);
}

currencyName =
myMoneyExchange.returnCurrencyName("France");
if (currencyName.equals("No match")) {
  System.out.println
  ("No Match Money Name Test PASSED "
  + currencyName);
} else {
  System.out.println("Swiss Money Name Test Failed "
  + currencyName);
}
//get equiv
returnedValue =
myMoneyExchange.returnForeignEquiv
("Argentina",1000);
if (returnedValue == 3755.0) {
  System.out.println("Converstion to Argentinian
  Money PASSED " + returnedValue);
} else {
    System.out.println("Converstion to Argentinian
    Money Failed " + returnedValue);
}

returnedValue =
myMoneyExchange.returnForeignEquiv
("Australia",1000);
if (returnedValue == 1830.0) {
  System.out.println("Converstion to Australian
  Money PASSED " + returnedValue);
```

```
    } else {
      System.out.println("Converstion to Australian
      Money Failed " + returnedValue);
    }

    returnedValue =
    myMoneyExchange.returnForeignEquiv("China",1000);

    if (returnedValue == 8290.0) {
      System.out.println("Converstion to Chinese Money
      PASSED " + returnedValue);
    } else {
      System.out.println("Converstion to Chinese Money
      Failed " + returnedValue);
    }

    returnedValue =
    myMoneyExchange.returnForeignEquiv("Japan",1000);
    if (returnedValue == 81.0) {
      System.out.println("Converstion to Japanese Money
      PASSED " + returnedValue);
    } else {
      System.out.println("Converstion to Japanese Money
      Failed " + returnedValue);
    }

    returnedValue =
    myMoneyExchange.returnForeignEquiv("Mexico",1000);
    if (returnedValue == 10120.0) {
      System.out.println("Converstion to Mexican Money
      PASSED " + returnedValue);
    } else {
      System.out.println("Converstion to Mexican Money
      Failed " + returnedValue);
    }

    returnedValue =
    myMoneyExchange.returnForeignEquiv("Swiss",1000);
    if (returnedValue == 1478.2) {
      System.out.println("Converstion to Swiss Money
      PASSED " + returnedValue);
    } else {
      System.out.println("Converstion to Swiss Money
```

```
          Failed " + returnedValue);
     }

     returnedValue =
     myMoneyExchange.returnForeignEquiv("UK",1000);

     if (returnedValue == 637.84) {
      System.out.println("Converstion to UK Money PASSED"
      + returnedValue);
     } else {
        System.out.println("Converstion to UK Money
        Failed " + returnedValue);
     }

     returnedValue =
     myMoneyExchange.returnForeignEquiv("France",1000);

     if (returnedValue == -1) {
      System.out.println("No match Test PASSED"
      + returnedValue);
     } else {
        System.out.println("Not match Test Failed "
        + returnedValue);
     }


     //repeat for other methods
      }
     }
```

When testing a Web Service or any other type of code, you want to not only test expected values but also what happens when you pass something that is out of range. You'll notice that there are two calls in the previous example that send France as the name of the country; the code, however, doesn't support this but still returns something to the method to let the program know. Supporting things that might cause an exception or out of range error are just as important to test as is seeing whether you receive the expected result.

This code is written as a command line application, but if you plan on supporting JSP and ASP.NET Web pages, you need to have tests that reside in those types of applications as well.

Testing methods and the different types of consumers are important, but so is knowing how many requests per second the server can handle. So by creating consumers that live in your supported consumers and then calling them through test software such as the Apache Group's JMeter™, you can gauge whether your server hardware will support your deployment.

THE DEPLOYMENT STAGE

By the time you reach the deployment stage, you should have an understanding of how reliable your code is, the number of requests that you can handle per second, and whether your hardware can support that demand. Usually, in the deployment stage, your customers start to give you their last-minute security demands, and you need to rush to implement them.

In this case, because the data isn't too personal, it doesn't need to be encrypted. However, you do still want to control who accesses the service. Sometimes it's difficult for people to find Web Services without something like UDDI available; but sometimes people do trip across your service and can access it simply by finding the WSDL. This is where a unique ID can help.

In the following code example there is a private method called `checkID`. This simple method contains a comparison between some string values that are assigned IDs. Then each method needs to have an ID parameter passed in to call the `checkID` static method. It was created as static so you would need to instantiate the `MoneyExchangeProtected` class just to call the `checkID` method. This way you can call it directly. If the proper ID is not passed, then either a numeric value indicates that the ID was incorrect or a text message is returned.

```
public class MoneyExchangeProtected {

  public double returnUSDollarEquiv
  (String ID, String countryName, float quantity) {

   double totalValue = 0;

   int idStatus =
       MoneyExchangeProtected.checkID(ID);

   if (idStatus == 0) {
```

```
        return -1;
    }

    //currency values as of 10/5/02
    if (countryName.equals("Argentina")) {
        totalValue = .2663 * quantity;
    } else if (countryName.equals("Australia")) {
        totalValue = .54 * quantity;
    } else if (countryName.equals("China")) {
     totalValue = .128 * quantity;
    } else if (countryName.equals("Japan")) {
     totalValue = .0081 * quantity;
    } else if (countryName.equals("Mexico")) {
     totalValue = .0988 * quantity;
    } else if (countryName.equals("Swiss")) {
     totalValue = .675 * quantity;
    } else if (countryName.equals("UK")) {
     totalValue = 1.568 * quantity;
    } else {
        totalValue = -1;
    }

    return totalValue;
 }

public double returnForeignEquiv
  (String ID, String countryName, float quantity) {
  double totalValue = 0;
  String name = null;

  if (countryName.equals("Argentina")) {
     name = "Peso";
  } else if (countryName.equals("Australia")) {
     name = "Dollar";
  } else if (countryName.equals("China")) {
     name = "Renminbi";
  } else if (countryName.equals("Japan")) {
     name = "Yen";
  } else if (countryName.equals("Mexico")) {
     name = "Peso";
  } else if (countryName.equals("Swiss")) {
     name = "Franc";
  } else if (countryName.equals("UK")) {
```

```
      name = "Pound";
   } else {
      name = "No match";
   }

   return name;

}


private static int checkID (String id) {
  int checkOK = 0;
  //ideally this calls a rdbms
  if (id.equals("bx113me") || id.equals("lx114me")) {
    checkOK = 1;
  }
  return checkOK;
}


}
```

This obviously isn't the only deployment issue, but by the time you reach this stage you should have gathered all your requirements, done preliminary testing, and completed more extensive testing including consumers and server strain. At this point, deployment should just be a matter of moving the files to the desired server and creating the appropriate security scheme.

CONCLUSION

Most of the Web Service examples presented in this book have been very simple in order to demonstrate the various technologies. In this chapter, you saw a closer examination of utilizing a Web Service by using an example that was more like something you would actually deploy.

This Money Exchange example demonstrated in this chapter allowed you to see how to pass various types of variables back and forth between the methods, and you saw a security example demonstrated in greater detail than in previous chapters of this book. Hopefully some of the information presented in this chapter will help you to deploy a real Web Service.

12

Using Web Services as a Middle Tier

Very often, the owners of a back-end system, such as a database, put a layer in between a Web site or other front-end system, including applications, to provide easy access to data. This way a Web site developer doesn't need to be concerned about the structure of a CRM or financial database. The developer focuses on how to call a particular object instead.

The objects and method calls remain static while the database or other data can constantly change. This may require changes to the methods themselves, but as long as the signatures remain the same, a Web developer need not change any Web pages.

Think of this in terms of a database-driven Web site. If there are hundreds of pages with SQL queries embedded in them and the database structure changes, each of the SQL statements residing in those Web pages will need to be changed. This becomes a huge maintenance issue, especially as a project grows and database changes need to take place to handle the large amounts of incoming data. With putting up a layer of objects, known as a middle tier, that contains methods to access a database between a Web site and the database, only a handful of SQL statements need to change within the objects. Very often these methods return some sort of results set or object with attributes, so the signature rarely changes.

It isn't just in accessing data in a *Remote Database Management System* (RDBMS) that a middle tier can act as a clearinghouse; it can also proxy requests to XML documents, *Excel*™ Files, CORBA objects, COM objects, or even other Web Services. If you support a middle tier like this, you provide your customers with a one-stop show for data, and they don't need to be concerned with multiple clients accessing different data types. They can simply interact with your middle tier, and there really isn't a better technology for a middle tier than a Web Service. Figure 12.1 demonstrates how Web Services may act as a middle tier.

Before Web Service became mainstream, it was very common to use either COM+ or CORBA as such a middle tier. If you have ever implemented one of these technologies, you understand the complexities of implementing and maintaining them. Both are fairly complex to implement, and it's very tough to find developers with the expertise to deploy these technologies in a reliable way. In addition, database developers usually are charged with implementing and maintaining a middle tier, and these team members rarely have the skills to implement COM+ or CORBA. That's the beauty of Web Services.

Web Services utilize Web servers and several different programming languages, and thus probably already fit into your infrastructure. Once a developer understands the basics of Web Services, they are easy to implement, especially because all

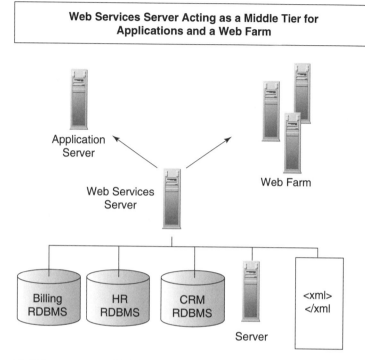

FIGURE 12.1 A middle tier utilizing Web Services can act as a clearinghouse to various data sources.

the hard work, such as forming the XML for a request, are abstracted in higher levels of code. With less training and infrastructure needs, a middle tier that utilizes Web Services can usually be set up fairly quickly and cheaply.

This chapter examines a middle tier that abstracts access to an XML file containing user information and eventually the Web Service discussed in Chapter 11.

CALLING THE XML

The first step in creating the middle tier example is creating the class needed to access the XML data. The example in this chapter utilizes a SAX-like parser to read the data as events. This provides a less efficient means of reading data, but provides fast access to the information. Even though this chapter examines accessing just one XML file, you could abstract access to several different files if you needed to.

The XML Data

The following XML data contains all the information needed for the three users allowed to access a particular Web site. Accessing data from XML will be faster than calling a RDBMS as long as you only have a few hundred or thousand records to access.

Notice that the root element is users and then the user parent tag represents each individual's data. The order of the elements usually doesn't matter for XML, but for the program shown later in the chapter the order does matter when it searches via the ID element. Remember that Chapter 2 provided a quick introduction to XML. You may want to refer to that chapter if the following XML doesn't make sense to you.

```xml
<?xml version='1.0'?>
<!-- This file represents a fragment of a book store
     inventory database -->
<users>
 <user>
   <id>101111</id>
   <password>hellodollie</password>
   <email>brianhochgurtel@hotmail.com</email>
   <title>Senior Programmer</title>
   <first-name>Brian</first-name>
   <last-name>Hochgurtel</last-name>

 </user>
 <user>
   <id>10112</id>
   <password>something</password>
   <email>secret@hotmail.com</email>
   <title>Senior Programmer</title>
   <first-name>David</first-name>
   <last-name>Pallai</last-name>
 </user>
 <user>
   <id>101113</id>
   <password>forever</password>
   <email>JohnDoe@hotmail.com</email>
```

```
      <title>Master Programmer</title>
      <first-name>John</first-name>
      <last-name>Doe</last-name>
   </user>
 </users>
```

This XML contains three different users, each with unique IDs and passwords. This makes it easy to use this information to restrict access to a Web site.

C# Code to Access XML

The following C# code sample uses the `XmlTextReader` class to grab data from the XML shown in the previous section. Because the data in this file is *element centric*, the only node you need to worry about in this case is `XMLNoteType.Text`. This provides you with the data that appears between the opening and closing tags.

The first step in this code defines which XML file to parse. In this example, the name of the file resides in a static string, but later examples actually provide a method to change the name and location of the XML file.

Then the `SearchStringXml` method defines a routine for searching the document for one particular string. You pass in a string, the document is parsed, and each value that appears between the elements is compared to the string passed in. If there is a match, 0 returns; otherwise, the return value will be 1.

This return value is the opposite of what you'd normally expect. A 1 should return for a condition that is true, but the string.Compare *method returns 0. This is something you should pay close attention to because it may cause some unexpected logic to occur in your code.*

The next method, `getData`, still parses only for the content of the elements, but now it does a search for the element you pass in. This code is set up to search by the value of the ID element. If you search on another element, the hash this class contains will return unexpected values. There are other parsing methods, such as DOM, that allow you to ask a parser for a particular element, but they are very memory intensive and will add time to accessing your data. This method of parsing the document is very fast but more inflexible.

Once the loop that searches for a match is true, a counter that keeps track of the number of elements is incremented each time. This allows the code to determine in which element to put the appropriate hash key. The count simply tracks which number of element you are on. For example, element Number 1 is ID, Number 2 contains the password, and Number 3 possesses the e-mail address. This count goes all the way up to 6 because there are 6 elements. This is why the order of the elements is so important in this case. If you move the elements around, the values end up with the incorrect hash key.

The main method then provides a means of testing the code you just created. The first call to the SearchStringXml method determines that the ID 101111 does exist within the XML. Then the call to the getData method puts all the appropriate personal information into the hash for the user with ID 101113.

```
using System;
using System.IO;
using System.Xml;
using System.Collections;

namespace XMLTester
{

  class GetXml
  {
    private static string myDocument = "users.xml";

    private int SearchStringXml
(String searchString)
      {
        int returnFlag = 0;
        XmlTextReader reader =
new XmlTextReader (myDocument);

        while (reader.Read())
        {
          switch (reader.NodeType)
          {
            case XmlNodeType.Text:
            returnFlag =
string.Compare(reader.Value,searchString);
            if (returnFlag == 1)
                return(1);

            break;
        }
        }
```

```
   return(0);
}

private Hashtable getData(String searchString)
{

XmlTextReader reader = new XmlTextReader (myDocument);
int returnFlag = 0;
 int elementCount =0;

Hashtable myHashTable = new Hashtable();

   while (reader.Read())
  {
    switch (reader.NodeType)
    {
       case XmlNodeType.Text:

        returnFlag =string.Compare (reader.Value,searchString);
        Console.WriteLine ("Debug:" + returnFlag +
          reader.Value + searchString);

        if (returnFlag == 0)
           elementCount++;

        if (elementCount >0 && elementCount <= 6)
        {
         if(elementCount == 1)
            {
            myHashTable.Add("id",reader.Value);
          elementCount++;
          Console.WriteLine ("Debug:" + reader.Value +
             elementCount);
          } else if(elementCount == 2) {

              myHashTable.Add("password",reader.Value);
            elementCount++;
            Console.WriteLine ("Debug:" + reader.Value +
               elementCount);
          } else if (elementCount == 3) {
              myHashTable.Add ("email",reader.Value);


            elementCount++;
```

```
                        Console.WriteLine ("Debug:" + reader.Value +
                            elementCount);
                } else if (elementCount == 4) {
                    myHashTable.Add("title",reader.Value);


                        elementCount++;
                        Console.WriteLine ("Debug:" +
                        reader.Value +
                            elementCount);
                } else if (elementCount == 5) {

myHashTable.Add
("FirstName",reader.Value);
                        elementCount++;
                        Console.WriteLine ("Debug:" + reader.Value +
                            elementCount);
                } else if (elementCount == 6) {
                    myHashTable.Add("LastName",reader.Value);
                        elementCount++;
                        Console.WriteLine ("Debug:" + reader.Value +
                            elementCount);
                }
            }
        break;
        }
    }
    return(myHashTable);
}


[STAThread]
static void Main(string[] args)
{
 GetXml myGetXml = new GetXml();

 try
 {
 // Load the file with an XmlTextReader
 int results = myGetXml.SearchStringXml("101111");

 Hashtable myHashTable = new Hashtable();
 myHashTable = myGetXml.getData("101113");

 Console.WriteLine ("Franklin was written:" + results);
```

```
Console.WriteLine ("ID:" + myHashTable["id"]);
Console.WriteLine ("email:" + myHashTable["email"]);
Console.WriteLine ("title:" + myHashTable["title"]);
Console.WriteLine ("firstname:" + myHashTable["FirstName"]);
Console.WriteLine ("lastname:" + myHashTable["LastName"]);
}
catch (Exception e)
{
  Console.WriteLine ("Exception: {0}", e.ToString());
 }


}
 }
}
```

Now that you know that the code works for the purposes of accessing XML data, the following code example takes the previous program and turns the code into a dll. This is completed by taking the code and removing the main, and then creating a project for a C# class library, as shown in Figure 12.2. This creates a dll that the Web Service example imports into its namespace.

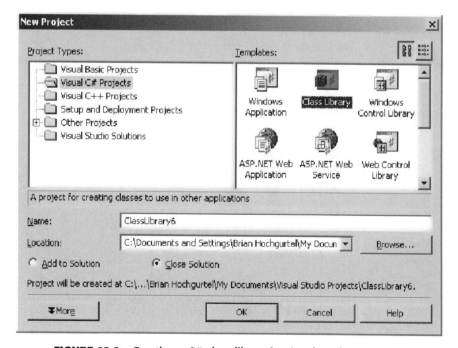

FIGURE 12.2 Creating a C# class library in *Visual Studio.NET*.

If you only have access to Microsoft's .NET SDK, you can compile the following code example into a dll using the command line C# compiler with the following command csc /t:library. *You also need to include any namespaces or other dlls needed for your library. Chapter 6 went into greater detail of using the command line compiler.*

Once you have your new project open, just paste in the following code and compile it. This creates a dll in your project directory that the Web Service application needs in order to call this XML parsing code.

```
using System;
    using System.IO;
    using System.Xml;
    using System.Collections;

    namespace XmlSearchString
    {
     class GetXmlLibrary
     {
      private string myDocument = "users.xml";

      public void setDocumentPath(string docPath)
      {
       myDocument = docPath;
      }


      public int SearchStringXml (String searchString)
      {
        int returnFlag = 0;
        XmlTextReader reader = new XmlTextReader (myDocument);

        while (reader.Read())
        {
        switch (reader.NodeType)
          {
           case XmlNodeType.Text:
           returnFlag = string.Compare(reader.Value,searchString);
          if (returnFlag == 1)
              return(1);

          break;
```

```
        }
      }
      return(0);
   }

   public Hashtable getData(String searchString)
   {

    XmlTextReader reader = new XmlTextReader
(myDocument);
      int returnFlag = 0;
      int elementCount =0;

    Hashtable myHashTable = new Hashtable();

    while (reader.Read())
    {
       switch (reader.NodeType)
          {
       case XmlNodeType.Text:
        returnFlag =
 string.Compare(reader.Value,searchString);


         if (returnFlag == 0)
            elementCount++;

            if (elementCount >0 && elementCount <= 6)
            {
                if(elementCount == 1)
                {
                myHashTable.Add("id",reader.Value);
elementCount++;

                }
                else if(elementCount == 2)
                {

myHashTable.Add("password",reader.Value);


                    elementCount++;
                }
                else if (elementCount == 3)
```

```
                       {

    myHashTable.Add ("email",reader.Value);

                          elementCount++;}
                 }
                 else if (elementCount == 4)
                 {
                     myHashTable.Add("title",reader.Value);
                     elementCount++;
                 }

                 else if (elementCount == 5)
                 {
                     myHashTable.Add("FirstName",reader.Value);
                     elementCount++;
                 }

                 else if (elementCount == 6)
                 {

                   myHashTable.Add("LastName",reader.Value);
                   elementCount++;
                 }

              }
           break;
            }
         }
        return(myHashTable);
      }
    }
  }
```

Once you are ready to deploy your dll, you should create a production build. This is done from the "Build" menu. Select "Build" and then "Configuration Manager" and you will see what appears in Figure 12.3.

On the "Configuration" drop-down box, select "Release." This removes all the debug symbols from your dll and makes the code execute faster.

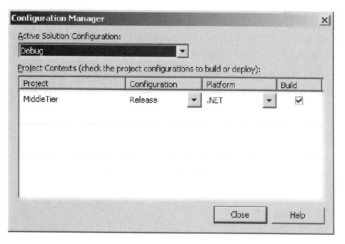

FIGURE 12.3   The "Build" settings for the XML parsing code.

MIDDLE TIER CODE

Now that we have the underlying XML code written, you can begin to write the code necessary to create the beginning of the middle tier. This section examines creating the code for the middle tier and then creating methods that make the most sense for utilizing Web Services.

Web Service with XML Access

The following code example contains C# code necessary to access the data in the XML file shown earlier in the chapter. The first method, validateId, compares the ID passed into the method to all the IDs in the XML file. If it finds a match using the string compare method, a 0 is returned—which seems opposite of what it should be.

The following method, ReturnEmail, returns the e-mail address of the person with the particular ID, and ReturnAll returns an array of all the values found in the XML file.

These methods in the Web Services are wrappers for the methods shown in the previous C# dll.

The XML methods return a hash rather than an array, but in order for the Web Service to return values, it must use an array or another structure that can be represented by the SOAP standard. It is worthwhile to run some tests with return values that are more complex than char *or* int. *Not all complex structures, such as hashes, can be represented by* SOAP.

```csharp
using System;
using System.Collections;
using System.ComponentModel;
using System.Data;
using System.Diagnostics;
using System.Web;
using System.Web.Services;
using XmlSearchString;


namespace MiddleTier
{
 [WebService(Description="XPlatform Middle Tier Example",
  Namespace="http://www.advocatemedia.com/")]

 public class Service1 : System.Web.Services.WebService
 {
      public Service1()
      {
          InitializeComponent();
      }

      #region Component Designer generated code

      //Required by the Web Services Designer
      private IContainer components = null;


      private void InitializeComponent()
      {
      }


      protected override void Dispose( bool disposing )
      {
```

```
            if(disposing && components != null)
            {
                  components.Dispose();
            }
            base.Dispose(disposing);
        }

        #endregion

        [WebMethod(Description="check if id exists")]

        public int ValidateId(string id)
        {
          GetXmlLibrary myGetXml =
new GetXmlLibrary();
          Hashtable myHashTable = new Hashtable();
          myHashTable = myGetXml.getData(id);
          string returnValue =
(string)myHashTable["id"];
          int result = string.Compare(id,returnValue);
          //weird -- 0 if true
          return(result);
        }


         [WebMethod
(Description="Return Email for Given id")]
        public string ReturnEmail(string id)
        {
          Hashtable myHashTable = new Hashtable();
          GetXmlLibrary myGetXml =
new GetXmlLibrary();
          myHashTable = myGetXml.getData(id);
          return((string)myHashTable["email"]);
        }

        [WebMethod(Description="Return Password for given id")]
        public string ReturnPassword(string id)
        {
          Hashtable myHashTable = new Hashtable();
          GetXmlLibrary myGetXml =
new GetXmlLibrary();
          myHashTable = myGetXml.getData(id);
          return((string)myHashTable["password"]);
```

```
        }

        [WebMethod(Description="Return all data as an array")]
        public string[] ReturnAll(string id)
        {
          Hashtable myHashTable = new Hashtable();
          GetXmlLibrary myGetXml =
new GetXmlLibrary();
          myHashTable = myGetXml.getData(id);
          string[] myStringArray = new string[6];
            myStringArray[0] =
(string)myHashTable["id"];
            myStringArray[1] =
(string)myHashTable["password"];
          myStringArray[2] =
(string)myHashTable["email"];
            myStringArray[3] =
(string)myHashTable["title"];
            myStringArray[4] =
(string)myHashTable["FirstName"];
            myStringArray[5] =
(string)myHashTable["LastName"];
          return(myStringArray);
        }

      }
    }
```

The Web Service acts like any other class. There are methods that return single important values and one method that returns everything. The question you need to ask yourself when deploying Web Services is if your design is efficient enough for the request and response model that HTTP and, therefore, Web Services are so dependent on.

Thinking about Efficient Web Services

This section examines how to make Web Services more efficient. Each method call requires a request and response to the server. Thus, if you are making several requests to a service for simple information, the response time of your application will be slower than for other technologies such as COM+.

For the previous example, take a look at the *SOAP* requests and responses to get an idea of how much traffic is really being generated.

First the application asks for the e-mail address for the particular ID.

```
<?xml version="1.0" encoding="utf-8"?>
<soap:Envelope
  xmlns:soap="http://schemas.xmlsoap.org/soap/envelope/"
  xmlns:xsi="http://www.w3.org/2001/XMLSchema-instance"
  xmlns:xsd="http://www.w3.org/2001/XMLSchema">
   <soap:Body>
     <ReturnEmail xmlns="http://www.advocatemedia.com/">
        <id>10112</id>
     </ReturnEmail>
   </soap:Body>
</soap:Envelope>
```

And the response is secret@hotmail.com.

```
Content-Type:text/xml; charset=utf-8

<?xml version="1.0" encoding="utf-8"?>
<soap:Envelope
 xmlns:soap="http://schemas.xmlsoap.org/soap/envelope/"
 xmlns:xsi="http://www.w3.org/2001/XMLSchema-instance"
 xmlns:xsd="http://www.w3.org/2001/XMLSchema">
   <soap:Body>
     <ReturnEmailResponse
       xmlns="http://www.advocatemedia.com/">
       <ReturnEmailResult>
           secret@hotmail.com
       </ReturnEmailResult>
     </ReturnEmailResponse>
   </soap:Body>
</soap:Envelope>
```

Now the method calls the service with the ID to return the password.

```
<?xml version="1.0" encoding="utf-8"?>
<soap:Envelope
 xmlns:soap="http://schemas.xmlsoap.org/soap/envelope/"
 xmlns:xsi="http://www.w3.org/2001/XMLSchema-instance"
 xmlns:xsd="http://www.w3.org/2001/XMLSchema">
 <soap:Body>
    <ReturnPassword
         xmlns="http://www.advocatemedia.com/">
```

```
        <id>10112</id>
      </ReturnPassword>
    </soap:Body>
  </soap:Envelope>
```

So the response comes back with the e-mail address.

```
Content-Type:text/xml; charset=utf-8

<?xml version="1.0" encoding="utf-8"?>
<soap:Envelope
 xmlns:soap="http://schemas.xmlsoap.org/soap/envelope/"
 xmlns:xsi="http://www.w3.org/2001/XMLSchema-instance"
 xmlns:xsd="http://www.w3.org/2001/XMLSchema">
  <soap:Body>
    <ReturnPasswordResponse
        xmlns="http://www.advocatemedia.com/">
      <ReturnPasswordResult>
        something
      </ReturnPasswordResult>
    </ReturnPasswordResponse>
  </soap:Body>
 </soap:Envelope>
```

Now that the information is confirmed for our individual user, let's return all the information about the user.

```
<?xml version="1.0" encoding="utf-8"?>
<soap:Envelope
xmlns:soap="http://schemas.xmlsoap.org/soap/envelope/"
xmlns:xsi="http://www.w3.org/2001/XMLSchema-instance"
xmlns:xsd="http://www.w3.org/2001/XMLSchema">
  <soap:Body>
    <ReturnAll xmlns="http://www.advocatemedia.com/">
      <id>10112</id>
    </ReturnAll>
  </soap:Body>
 </soap:Envelope>
```

Now, the following is the long response containing all the information about this individual.

```
Content-Type:text/xml; charset=utf-8

<?xml version="1.0" encoding="utf-8"?>
<soap:Envelope
xmlns:soap="http://schemas.xmlsoap.org/soap/envelope/"
xmlns:xsi="http://www.w3.org/2001/XMLSchema-instance"
xmlns:xsd="http://www.w3.org/2001/XMLSchema">
  <soap:Body>
    <ReturnAllResponse
      xmlns="http://www.advocatemedia.com/">
      <ReturnAllResult>
        <id>10112</id>
        <password>something</password>
        <email>secret@hotmail.com</email>
        <title>Senior Programmer</title>
        <FirstName>David</FirstName>
        <LastName>Pallai</LastName>
        <returnAmountUs>0</returnAmountUs>
        <returnAmountForeign>0</returnAmountForeign>
      </ReturnAllResult>
    </ReturnAllResponse>
  </soap:Body>
</soap:Envelope>
```

The transactions shown in the previous *SOAP* messages demonstrate a typical exchange between a Web page and a database server, but for that type of connection these types of transactions make sense. Many times the connection between a Web server and a database is held open—making the transactions very fast. With Web Services, the exchanges take more time because the connection cannot be held open, because the HTTP standard doesn't allow for that. So you need to examine the structure and efficiency of your methods and how you access them. For example, in the request and response shown previously, it would have been much faster just to return all the information first rather than confirming each piece.

Because a Web Service does generate so much Web Service traffic, it is better to design methods that contain everything you need to do within the Web Service. That may mean cramming several method calls to different objects within one Web method, and this probably violates good object oriented design. But it does cut down on the amount of traffic a Web Service generates.

Let's make the Web Service generate even more traffic by adding one more layer to the middle tier by creating a proxy to the Web Service shown in Chapter 11.

Adding a Proxy to `MoneyExchangeService`

The following code snippet creates proxy methods to each of the methods in the Java Web Service `ReturnCurrencyName`. By adding a Web reference to this service in *Visual Studio*, we can then wrap each method as a C# Web method. Notice how each method creates an instance of the object, calls the method, and then returns the value. C# does very little work here.

```
[WebMethod(Description="Get the name of the currency")]
 public string ReturnCurrencyName(string countryName)
 {
  advocatemedia.MoneyExchangeService myExchange = new
  advocatemedia.MoneyExchangeService();
  string moneyName = myExchange.returnCurrencyName(countryName);
 return(moneyName);
 }

[WebMethod(Description="Foreign Equivalent of US Dollars")]
 public double ReturnForeignEquiv(string country, float amount)
 {
 advocatemedia.MoneyExchangeService myExchange = new
  advocatemedia.MoneyExchangeService();
 double returnedAmount =
  myExchange.returnForeignEquiv(country, amount);
  return(returnedAmount);
 }

[WebMethod(Description="Return the US Dollar Equiv ")]
 public double ReturnUSDollarEquiv(string country, float amount)
 {
 advocatemedia.MoneyExchangeService myExchange = new
  advocatemedia.MoneyExchangeService();
 double returnedAmount =
  myExchange.returnUSDollarEquiv(country, amount);
  return(returnedAmount);
 }
```

This adds *even more* traffic because every request to this service causes a request to be made to another service. So not only is there a request and response to this service but also one to `MoneyExchangeService`. Figure 12.4 demonstrates this.

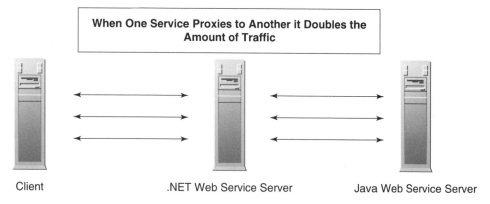

FIGURE 12.4   Proxy requests to another Web Service doubles the amount of traffic being generated.

The next section looks at ways of making Web Services more efficient.

MAKING WEB SERVICES MORE EFFICIENT

This section looks at two ways of making Web services more efficient. The first involves creating another class within the service to act as a container for values to be returned. The second looks at consolidating several method calls into one Web method to reduce network traffic.

Creating Another Class

A class who has no methods and all public variables provides for a handy container for passing back values to the consumer. This guarantees that you use a data structure that can be represented in *SOAP* without error. Consider the following example.

```
public class ReturnValues
 {
  public string id;
  public string password;
  public string email;
  public string title;
  public string FirstName;
```

```
public string LastName;
public string moneyName;
public double returnAmountUs;
public double returnAmountForeign;
}
```

Now there is a container for storing the data that can be easily accessed and passed back. Because all the variables are public there is no need for having accessor methods.

By putting this class in a Web Service, it creates the following construct in the WSDL. Note there is no means to serialize methods, only attribute variables of a class. This needs to be in the WSDL so the proxy, when generated, knows how to handle this user-defined type.

```
<s:complexType name="ReturnValues">
<s:sequence>
 <s:element minOccurs="0" maxOccurs="1" name="id" type="s:string" />
 <s:element minOccurs="0" maxOccurs="1" name="password" type="s:string"
/>
 <s:element minOccurs="0" maxOccurs="1" name="email" type="s:string" />
 <s:element minOccurs="0" maxOccurs="1" name="title" type="s:string" />
 <s:element minOccurs="0" maxOccurs="1" name="FirstName"
type="s:string" />
 <s:element minOccurs="0" maxOccurs="1" name="LastName" type="s:string"
/>
 <s:element minOccurs="0" maxOccurs="1" name="moneyName"
type="s:string" />
 <s:element minOccurs="1" maxOccurs="1" name="returnAmountUs"
type="s:float" />
 <s:element minOccurs="1" maxOccurs="1" name="returnAmountForeign"
type="s:float" />
 </s:sequence>
      </s:complexType>
```

This WSDL code then generates the following code in the proxy.

```
[System.Xml.Serialization.XmlTypeAttribute
(Namespace="http://www.advocatemedia.com/")]

public class ReturnValues {

    /// <remarks/>
```

```
        public string id;

        /// <remarks/>
        public string password;

        /// <remarks/>
        public string email;

        /// <remarks/>
        public string title;

        /// <remarks/>
        public string FirstName;

        /// <remarks/>
        public string LastName;

        /// <remarks/>
        public string moneyName;

        /// <remarks/>
        public System.Single returnAmountUs;

      /// <remarks/>
    public System.Single returnAmountForeign;
}
```

Then, when you call a Web Service that contains this class, you must create it before you pass any values to it. In the following code snippet, myReturnValues is an object of the ReturnValues type. Notice that it is very easy to display data from this object in the Console.WriteLine statement.

```
ReturnValues myReturnValues = new ReturnValues();
Service1 myService = new Service1();

myReturnValues = myService.testReturn();

Console.WriteLine("This is the email " +
myReturnValues.email);
```

This creates an handy container for passing values back and forth to the Web Service.

Making Methods Efficient

The point of making methods more efficient is to reduce the amount of traffic occurring on the network. The best way to do that is to encapsulate a unit of work within the method. This means that you put everything you need to do with one Web Service inside one method. Consider the doAll method in the following code snippet.

```
[WebMethod
(Description="Get information from ID and return money
 info")]

public ReturnValues doAll(string id, string country,
float usDollars, float foreignDollars)
{
  ReturnValues myReturnValues = new ReturnValues();
  myReturnValues.moneyName =
  myExchange.returnCurrencyName(country);
  myReturnValues.returnAmountUs =
  myExchange.returnUSDollarEquiv
  (country, foreignDollars);
  myReturnValues.returnAmountForeign =
  myExchange.returnForeignEquiv(country, usDollars);

  Hashtable myHashTable = new Hashtable();
  GetXmlLibrary myGetXml = new GetXmlLibrary();
  myHashTable = myGetXml.getData(id);


  myReturnValues.id = (string)myHashTable["id"];

   myReturnValues.password =
  (string)myHashTable["password"];

  myReturnValues.email    = (
  (string)myHashTable["email"];

  myReturnValues.title    =
  (string)myHashTable["title"];

   myReturnValues.FirstName =
  (string)myHashTable["FirstName"];
```

```
    myReturnValues.LastName =
    (string)myHashTable["LastName"];

    return(myReturnValues);
}
```

The previous method takes all possible methods from the Web Service and puts them in the same method and passes all the information back to the consumer. This reduces the number of Web requests and response by two, which Figure 12.5 demonstrates.

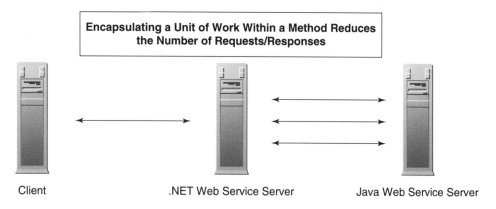

FIGURE 12.5 Putting the unit of work within a single method reduces the amount of traffic.

Or if that's too much, there's the doAllMoney method in the following code snippet that only works on exchanging money.

```
[WebMethod(Description="Do all money related calls")]
public ReturnValues doAllMoney
(string country, float usDollars, float foreignDollars)
{
 ReturnValues myReturnValues = new ReturnValues();
 myReturnValues.moneyName =
  myExchange.returnCurrencyName(country);

 myReturnValues.returnAmountUs =
```

```
        myExchange.returnUSDollarEquiv(country, foreignDollars);

        myReturnValues.returnAmountForeign =
        myExchange.returnForeignEquiv(country, usDollars);

        return(myReturnValues);
    }
```

The following code example is the complete modified Web Service so you can see the placement of the ReturnValues class.

The hash used in the XML methods could have been an object too, but it is used here to show that you need to test the types you return from objects in libraries to ensure they can be passed back by Web Services.

```
using System;
    using System.Collections;
    using System.ComponentModel;
    using System.Data;
    using System.Diagnostics;
    using System.Web;
    using System.Web.Services;
    using XmlSearchString;


    namespace MiddleTier
    {
    public class ReturnValues
    {
     public string id;
     public string password;
     public string email;
     public string title;
     public string FirstName;
     public string LastName;
     public string moneyName;
     public double returnAmountUs;
     public double returnAmountForeign;

    }
```

```
[WebService(Description="XPlatform Middle Tier Example",
 Namespace="http://www.advocatemedia.com/")]

public class Service1 : System.Web.Services.WebService
{
    public Service1()
    {
     InitializeComponent();
    }

      #region Component Designer generated code

      //Required by the Web Services Designer
      private IContainer components = null;

      private void InitializeComponent()
      {
      }

      protected override void Dispose( bool disposing )
      {
         if(disposing && components != null)
         {
          components.Dispose();
         }
        base.Dispose(disposing);
      }

      #endregion
      private advocatemedia.MoneyExchangeService myExchange =
        new advocatemedia.MoneyExchangeService();

      [WebMethod(Description="check if id exists")]
      public int ValidateId(string id)
      {
        GetXmlLibrary myGetXml =
             new GetXmlLibrary();
        Hashtable myHashTable = new Hashtable();
        myHashTable = myGetXml.getData(id);
        string returnValue =
             (string)myHashTable["id"];
        int result =
             string.Compare(id,returnValue);
```

```
            //weird -- 0 if true
            return(result);
        }


<DIS[WebMethod
        (Description="Return Email for Give Address")]
        public string ReturnEmail(string id)
        {
          Hashtable myHashTable = new Hashtable();
          GetXmlLibrary myGetXml =
              new GetXmlLibrary();
          myHashTable = myGetXml.getData(id);
          return((string)myHashTable["email"]);
        }


        [WebMethod(Description="Return Password for given id")]
        public string ReturnPassword(string id)
        {
          Hashtable myHashTable = new Hashtable();
          GetXmlLibrary myGetXml =
              new GetXmlLibrary();
          myHashTable = myGetXml.getData(id);
          return((string)myHashTable["password"]);
        }


        [WebMethod
            (Description="Return all data as an array")]
        public ReturnValues ReturnAll(string id)
        {
          Hashtable myHashTable = new Hashtable();
          GetXmlLibrary myGetXml =
              new GetXmlLibrary();
          myHashTable = myGetXml.getData(id);

            ReturnValues myReturnValues =
                new ReturnValues();
            myReturnValues.id       =
                (string)myHashTable["id"];
            myReturnValues.password =
            (string)myHashTable["password"];
            myReturnValues.email    =
                (string)myHashTable["email"];
            myReturnValues.title    =
```

```
            (string)myHashTable["title"];
      myReturnValues.FirstName =
        (string)myHashTable["FirstName"];
        myReturnValues.LastName =
        (string)myHashTable["LastName"];

    return(myReturnValues);
  }

[WebMethod
      (Description="Get the name of the currency")]
    public string ReturnCurrencyName
        (string countryName)
    {
      string moneyName =
      myExchange.returnCurrencyName(countryName);
    return(moneyName);
    }

[WebMethod
      (Description="Return the Foreign Equivalent")]
    public double ReturnForeignEquiv
        (string country, float amount)
    {
     double returnedAmount =
       myExchange.returnForeignEquiv
           (country, amount);
       return(returnedAmount);
    }

    [WebMethod
        (Description="Return the US Dollar Equiv ")]
    public double ReturnUSDollarEquiv
      (string country, float amount)
    {
     double returnedAmount =
       myExchange.returnUSDollarEquiv
           (country, amount);
     return(returnedAmount);
    }

    [WebMethod
        (Description="Get information and return money")]
```

```csharp
public ReturnValues doAll
(string id, string country, float usDollars,
   float foreignDollars)
{
  ReturnValues myReturnValues =
       new ReturnValues();
    myReturnValues.moneyName =
    myExchange.returnCurrencyName(country);
  myReturnValues.returnAmountUs =
    myExchange.returnUSDollarEquiv
        (country, foreignDollars);
    myReturnValues.returnAmountForeign =
    myExchange.returnForeignEquiv
        (country, usDollars);

  Hashtable myHashTable = new Hashtable();
  GetXmlLibrary myGetXml =
       new GetXmlLibrary();
  myHashTable = myGetXml.getData(id);


  myReturnValues.id        =
      (string)myHashTable["id"];
  myReturnValues.password =
    (string)myHashTable["password"];
  myReturnValues.email     =
      (string)myHashTable["email"];
  myReturnValues.title     =
      (string)myHashTable["title"];

    myReturnValues.FirstName =
    string)myHashTable["FirstName"];

    myReturnValues.LastName =
    (string)myHashTable["LastName"];

  return(myReturnValues);
}

[WebMethod
    (Description="Do all money related calls")]
public ReturnValues doAllMoney
```

```
    (string country, float usDollars, float foreignDollars)
    {
    ReturnValues myReturnValues = new ReturnValues();
    myReturnValues.moneyName =
      myExchange.returnCurrencyName(country);
    myReturnValues.returnAmountUs =
      myExchange.returnUSDollarEquiv(country, foreignDollars);
    myReturnValues.returnAmountForeign =
      myExchange.returnForeignEquiv(country, usDollars);

    return(myReturnValues);
    }
  }
}
```

Methods that do many things are excellent for encapsulating a unit of work with a Web Service.

TIP

Another thing to consider with a Web Service is what you are using it for. For example, you would not want to use a Web service for session management or to keep track of state (logged in/logged out) on a Web site. That requires too much traffic, especially when you constantly need to communicate with the database. Web Services can be used for setting states that are cookie based, this is what the following set of examples shows.

USING THE MIDDLE TIER WEB SERVICE TO TRACK STATE

This section discusses *ASP.NET* Web pages that communicate with the Middle Tier Web Service shown earlier in the chapter. This Web page communicates with the service and uses the values passed back to set an authenticated cookie. This way the Web pages can look for the cookie to see if the user is authorized, rather than having to communicate with the Web Service.

Within a *.NET* directory that requires form-based security, two files must be present, `login.aspx` and `default.aspx`, for this authentication process to work. In addition, modifications need to be made to the `web.config file`.

Setting up `Web.config`

When you create a *ASP.NET* project for C# in *Visual Studio.NET*, it generates an XML-based file called `Web.config`. When you open this file, you'll notice that there are several directives. The following XML shows the different elements that need to be added between the `system.web` elements.

The authentication element's `mode` attribute set to `Forms` means to use forms-based authentication rather than using *Windows* or network settings. Within the authorization elements, `deny users = "?"` tells the *ASP.NET* application to deny anonymous users.

```
<system.web>
  <authentication mode="Forms"/>
  <authorization>
    <deny users="?" />
  </authorization>
</system.web>
```

Creating `Login.aspx`

The following code sample is an *ASP.NET* Web page required to appear in a directory that is protected by form-based authentication. Each page in this directory contains the `<%@ Import Namespace="System.Web.Security" %>` directive, which not only provides access to objects needed for security but also forces each page to examine how security is maintained and then to check for that security method. If someone hasn't completed a login, the page routes the individual to `login.aspx` and asks for their credentials. The next import statement makes all the methods from the Web Service available.

Then the HTML begins with defining code within the script tags. This contains one method, `Login_Click`, which examines the values of the user ID and password sent to the form compared to those found in the XML file. The first step involves creating a `ReturnValues` object so that the appropriate values are returned from the service. The next step involves creating the object to communicate with the Web Service and then passing the user ID submitted to the form. The information returns all the information, *which is probably most efficient for the Web Service*, and then the comparison occurs between the XML data and the data in the form.

If the comparison is true, the `FormsAuthentication.RedirectFromLoginPage` method is called to create the authentication cookie. This first parameter is the variable you wish to use as a unique ID to track the user. Many people use either an e-mail address or a unique number, as in this case. The Boolean, `myCookieSetting`, tells the method to use a cookie to track the user.

After the script tags, the HTML form is defined. This uses *.NET* elements seen in other *ASP.NET* examples shown in previous chapters, but the `ASP:Required-FieldValidator` element is new. Both of these ASP directives tell the page that values must be entered for the user ID and password fields.

```
<%@ Page language="c#"
         Codebehind="WebForm1.aspx.cs"
         AutoEventWireup="false"
         Inherits="SecurityExample.WebForm1" %>
<%@ Import Namespace="System.Web.Security" %>
<%@ Import Namespace="SecurityExample.localhost" %>

<html>

  <script language="C#" runat=server>

    void Login_Click(Object sender, EventArgs E) {
      ReturnValues myReturnValues = new ReturnValues();
      Service1 myService = new Service1();
      myReturnValues =
      myService.ReturnAll(UserId.Value);

      int idResult =
      string.Compare(UserId.Value,myReturnValues.id);

      int passwordResult = string.Compare
      (UserPass.Value,myReturnValues.password);

      bool myCookieSetting = true;

      if (idResult == passwordResult) {
        FormsAuthentication.RedirectFromLoginPage
        (myReturnValues.id, myCookieSetting);
      } else {
        Msg.Text = "Invalid password, -try again";
      }
    }

  </script>
```

```html
<body>

  <form runat=server ID="Form1">

   <h3><font face="Verdana">Login Page</font></h3>

   <table>
     <tr>
     <td>UserId:</td>
     <td>
      <input id="UserId" type="text"
             runat=server NAME="UserId">
     </td>
     <td><ASP:RequiredFieldValidator
            ControlToValidate="UserId"
            Display="Static"
            ErrorMessage="*"
            runat=server
            ID="myRequiredfieldvalidator1"
            NAME="myRequiredfieldvalidator1"/>
     </td>
   </tr>
   <tr>
     <td>Password:</td>
     <td><input id="UserPass"
               type=password
               runat=server
               NAME="UserPass"/>
     </td>
     <td><ASP:RequiredFieldValidator
            ControlToValidate="UserPass"
            Display="Static" ErrorMessage="*"
            runat=server
            ID="myRequiredfieldvalidator2"
            NAME="myRequiredfieldvalidator2"/>
     </td>
   </tr>

   </table>

     <asp:button text="Login" OnClick="Login_Click"
                runat=server ID="Button1"
                NAME="Button1"/>
```

```
    <p>

      <asp:Label id="Msg" ForeColor="red"
                 Font-Name="Verdana" Font-Size="10"
                 runat=server />

    </form>
    </body>

  </html>
```

Figure 12.6 shows Login.aspx in *Internet Explorer*.

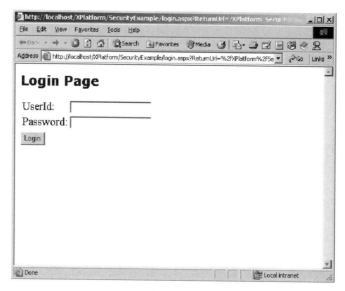

FIGURE 12.6   Login.aspx in *Internet Explorer*.

Creating Default.aspx

Default.aspx is the equivalent to index.html for other Web applications. It's the page the application sends users to when they first come to the directory where the *ASP.NET* pages reside. Notice in the following code example that this page has the same import statements that the previous example did. This causes the page to check for the security credentials created in the previous example; if it doesn't find them, the user is sent back to `login.aspx`. The includes also provide access to the methods in the middle tier Web Service.

Within the script elements reside two methods. The first one, `Signout_Click`, uses the `SignOut()` method to delete the user's authorization cookie and then uses the `Response` object's `Redirect` method to send the user back to `login.aspx`.

The second method calls the middle tier Web Service created earlier in the chapter. This method uses values sent in from the Web form to call the Web Service and send values back to the appropriate *ASP.NET* controls on the Web page.

TIP

Using cookie-based authentication prevented this example from making another call to a Web Service to ensure the user is logged in. This is a great way to go when utilizing Web Services because it cuts down on network traffic because the login state is stored on the client.

```
<%@ Page language="c#"
    Codebehind="default.aspx.cs"
    AutoEventWireup="false"
    Inherits="SecurityExample._default" %>
<%@ Import Namespace="System.Web.Security " %>
<%@ Import Namespace="SecurityExample.localhost" %>


<html>

  <script language="C#" runat=server>

  void Signout_Click(Object sender, EventArgs E) {

    FormsAuthentication.SignOut();
    Response.Redirect("login.aspx");
  }

  void Submit_Click(Object sender, EventArgs E) {
    ReturnValues myReturnValues =
    new ReturnValues();
    Service1 myService = new Service1();
    float myDollars =
    float.Parse(USDollars.Text);
```

```
     float myForeignDollars =
     float.Parse(ForeignDollars.Text);

     string myCountryName =
     countryName.SelectedItem.Text;

     myReturnValues = myService.doAllMoney
     (myCountryName, myDollars, myForeignDollars);


     foreign.Text =
     "The foreign amount returned was"
     + myReturnValues.returnAmountForeign;

     us.Text = "Amount in American Dollars: "
     + myReturnValues.returnAmountUs;

     moneyname.Text = myReturnValues.moneyName;
}

</script>

<body>

 <h3>Do Money Conversion</h3>

<form runat=server ID="doConversions">
 <h3><asp:label id="Welcome" runat=server/></h3>
     <asp:label id="foreign" runat=server/><br>
     <asp:label id="us" runat=server/><br>
     <asp:Label ID="moneyname" Runat=server/><br>



<hr>
 Select a country:
 <asp:ListBox id = "countryName" Runat=server>
   <asp:ListItem></asp:ListItem>
   <asp:ListItem>Argentina</asp:ListItem>
   <asp:ListItem>Australia</asp:ListItem>
   <asp:ListItem>China</asp:ListItem>
   <asp:ListItem>Mexico</asp:ListItem>
   <asp:ListItem>Swiss</asp:ListItem>
```

```
        <asp:ListItem>UK</asp:ListItem>
    </asp:ListBox>

    <br>
     Enter amount in US Dollars:
     <asp:TextBox ID="USDollars" Runat=server/>
    <br>
     Enter amount in Foreign Currency:
    <asp:TextBox ID="ForeignDollars" Runat=server/>
    <br>
    <input type="submit"
           value="get money name"
           onserverclick="Submit_Click"
           runat="server"
           ID="Submit1"
           NAME="Submit1"/>

    <HR width=100% size=1>
    <asp:button text="Signout"
                OnClick="Signout_Click"
                runat=server
                ID="Button1"
                NAME="Button1"/>

    </form>
   </body>
 </html>
```

Figure 12.7 shows Default.aspx in *Internet Explorer.*

From both Web page examples in this section, the methods called from the Web Service look like any other methods. The developer creating these pages has no idea that this data comes from an XML file.

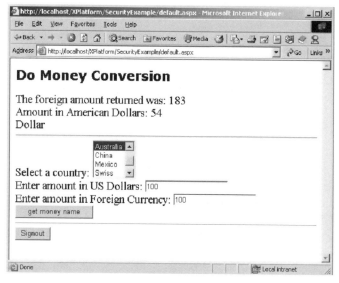

FIGURE 12.7   `Default.aspx` in *Internet Explorer*.

CONCLUSION

This chapter demonstrates how a Web Service acts as a middle tier to Web pages hiding and abstracting the data source from the Web page or application developer. This is useful because the method and data source can change without affecting what could be thousands of consumers.

Another advantage to using Web Services is that it is likely that you already have the hardware in place to deploy them. It's far more likely that your people possess the skills to deploy the technology. This may not be the case when using COM+ or CORBA as a middle tier.

13

Creating Your Own Web Services Implementation

In This Chapter

- Making the HTTP Request
- Creating the *SOAP* Request
- Reading the Request
- Server Implementation and WSDL
- Communicating with Web Services in a Nonstandard Way

I f you can't find a Web Services implementation for the platform and/or language you wish to support, you may need to look at writing your own Web Services implementation. This is a large undertaking that involves considering the many standards that Web Services revolve around including HTTP, XML, WSDL, UDDI, and *SOAP*.

Beyond the standards, you need to consider whether your Web Services will be server-side based or client-side based only. Many implementations of Web Services are simply client-side so that the application can communicate with the outside world. This, however, doesn't work well when the outside world needs to call your legacy system. This leads to creating something that serves Web Services, but then you also need to consider with what systems your implementation will be compatible.

Another alternative is to communicate with Web Services in a nonstandard way. This means that you use a Get or Post method to the Web Service and capture the results. Then utilize whatever string or XML parsing method you have available to you.

MAKING THE HTTP REQUEST

The first step of creating a Web Service implementation is having a library or piece of code that can make a request to a URL out on the Internet. This involves following the HTTP standard that the W3C supports. The best bet with this piece is to find a pre-existing piece of code or a library that can make this call for you.

Almost all languages, such as Java or C#, have the ability to call a URL. A language that doesn't handle this is C++, but you can purchase a library to make such a call. If you have to create the code to make a HTTP request, this process is going to take a great deal of time. Figure 13.1 demonstrates that making the HTTP request is just the first step in creating a request to a Web Service.

CREATING THE *SOAP* REQUEST

In this step, you must put the *SOAP* XML document into the HTTP request. This means you have to use elements in the *SOAP* standard for the server to be able to

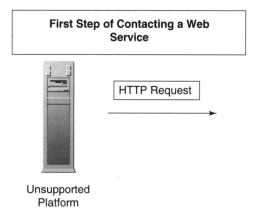

FIGURE 13.1 Making the HTTP request is the first step in creating a Web Service implementation.

parse the document and provide the appropriate response. Assembling an XML document within a program does not require a parser. Consider the following code snippet.

```
String BodyVariable1 =
"<countryNamexsi:type=\"xsd:string\">";
```

This simple piece of code starts to build an XML document. The problem here is that there is no way to determine whether the XML you build is well formed and valid. Thus, it may be difficult to determine why your request fails to the server. If an XML parser is unavailable to you, creating a method that can dump out the XML your program creates gives you the opportunity to move the XML document to a place where you can test the document's validity. Even just trying to load this document into *Internet Explorer* would be sufficient. If a parser such as Apache *Xerces* is available to you, a Web Services implementation is probably also available. Figure 13.2 shows that creating the request involves piggybacking the XML with the HTTP request.

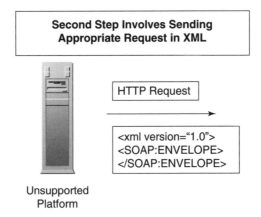

FIGURE 13.2   The HTTP request to the server
must include the appropriate XML from the
SOAP standard.

READING THE REQUEST

It would be fairly easy to assemble an XML document to make a request, but when
the response comes back you need to have a way to parse. An XML parser can
server the document up in its various pieces; however, if that functionality is not
available to you there needs to be some way to break the document up into its appropriate pieces. Otherwise your program will have the ability to send the request
without the ability of reading the response.

It would be possible to break up the document using a series of substring and
other string-parsing methods, but the location of the elements can vary inside the
document. Thus, there is no guarantee this method would work. Figure 13.3 illustrates how reading the response is the next piece of the puzzle.

SERVER IMPLEMENTATION AND WSDL

Will your implementation be able to serve the Web Services you create? If so, your
code will need to be able to send and receive HTTP requests. Your executable will
need to bind to a particular port and listen for requests. The XML that comes in

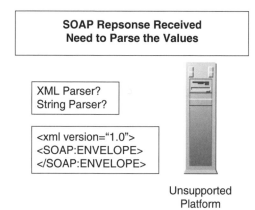

FIGURE 13.3 Reading the response is the next part of designing your Web Services software.

will need to be routed to the appropriate executable and parsed to get to its information. Then the appropriate response will need to be assembled and sent back to the client.

Many times developers create clients so that a particular environment has access to the outside world. For example, the Apache group will be releasing a C++ client for *Axis* sometime soon. It only works on the client side because Java can access C++ through the *Java Native Interface* (JNI) from an *Axis* service. The client is useful if there is information that C++ needs from Java. It's more difficult to go in the other direction.

For your server code to be useful, you need to consider how to describe it to clients. If you don't involve the WSDL standard, you may find yourself having to create a specialized client that is not available to standard implementations such as *.NET* and *Axis*. Even though it may appear that using WSDL is the easy way to compatibility with other implementations, it is still possible to create a perfectly valid WSDL document that *.NET* and *Axis* would be unable to read. That's just a natural danger with XML; it is so loosely structured by design that two documents can follow the same rules and still be incompatible.

It is not a requirement that your Web Service generate WSDL. However, if WSDL is not generated, it is more difficult for your Web Service to communicate with other distributions. In addition, clients that don't require WSDL will need to have more lines of code to call your service. Thus, it becomes somewhat inconvenient for the developer.

Another note to consider with WSDL is that many Web Service implementations didn't catch on until WSDL came in and the proxy was created, thus making it much easier to instantiate a Web Service. Therefore, WSDL is a feature you definitely want to employ with any Web Service tools you create. Figure 13.4 shows how acting as a server opens Web Service software possibilities.

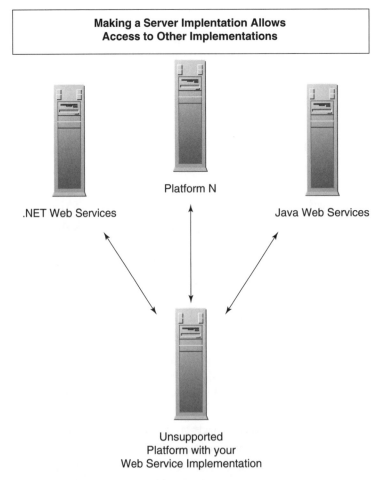

FIGURE 13.4 How implementing Web Services software as a server implementation opens access to other platforms.

COMMUNICATING WITH WEB SERVICES IN A NONSTANDARD WAY

If you do not have a Web Services client for your platform and you still wish to communicate with a Web Service, this is possible to do in a nonstandard way. This means that you ignore many of the standards mentioned in this book, and use both HTTP Post and Get methods to send information to your application. This section addresses this topic using a Java example and some HTML forms.

Making a Request to an *Axis* Web Service with HTML

Both *Axis* and *.NET* Web Services respond to Post and Get requests. Using an HTML form with *Axis* will demonstrate this. Consider the following HTML.

```
<HTML>
<HEAD>
    <TITLE>Test Web Service</TITLE>
</HEAD>

<FORM METHOD="GET"
 ACTION="http://localhost:8080/axis/MoneyExchange.jws">

 Method Name:
 <input type="text"
        name="method"
        value="returnCurrencyName">
 <BR>
 Country Name:
 <input type="text"
        name="countryName"
        value="Argentina">

 <input type="submit">

</FORM>
</HTML>
```

Figure 13.5 displays the HTML form within *Internet Explorer*.

The HTML makes a Get request to the MoneyExchange Web Service described in Chapter 11. The default values of the HTML text inputs call the returnCurrencyName method with a default value of the countryName being Argentina. When you load the

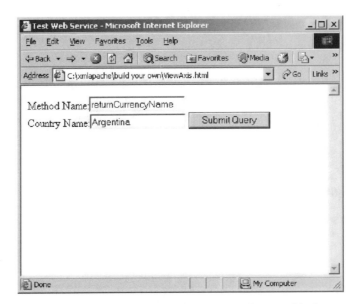

FIGURE 13.5 The HTML form that communicates with the `MoneyExchange` *Axis* Web Service.

form into *Internet Explorer* and submit it, you'll get the following response for the default values.

```
<p>Got response message</p>
<?xml version="1.0" encoding="UTF-8"?>
<soapenv:Envelope
 xmlns:soapenv=
"http://schemas.xmlsoap.org/soap/envelope/"
 xmlns:xsd=
"http://www.w3.org/2001/XMLSchema"
 xmlns:xsi=
 "http://www.w3.org/2001/XMLSchema-instance">
 <soapenv:Body>
 <returnCurrencyNameResponse soapenv:encodingStyle=
 "http://schemas.xmlsoap.org/soap/encoding/">
    <returnCurrencyNameReturn xsi:type="xsd:string">
    Peso
```

```
    </returnCurrencyNameReturn>
   </returnCurrencyNameResponse>
  </soapenv:Body>
 </soapenv:Envelope>
```

This is the expected *SOAP* response from the Web Service. By using a `Get` method from a HTML form we were able to communicate with the service.

When this XML comes back to Internet Explorer, *you'll get an error stating that you can't have two root elements. This happens because executing a* `Get` *against an* Axis *Web Service, returns* `<p>Got response message</p>` *at the beginning of the request. This prevents* Internet Explorer *from parsing the XML correctly. To re-trieve the XML, do a "View Source" and* Internet Explorer *loads the XML into* Notepad, *and then you can copy and paste it anywhere you'd like.*

If you communicate with an Axis *Web Service using the* `Post` *method, it expects to receive a complete* SOAP *request and this just won't work within an HTML form.*

This example shows how to communicate with *Axis* in a rudimentary fashion. The next section demonstrates how to accomplish the same thing with *.NET*.

Making a Request to an *.NET* Web Service with HTML

The previous example demonstrated how to communicate with an *Axis* Web Service. The following HTML communicates with the `GetTestQuote` *.NET* Web Service demonstrated in Chapter 6. Note that a *.NET* Web Service will communicate through an HTML form either with a `Post` or a `Get` request.

```
<HTML>
<HEAD>
   <TITLE>Test Web Service</TITLE>
</HEAD>

<FORM
 METHOD="POST" ACTION=
"http://www.advocatemedia.com/webservices/service1.asmx
/GetTestQuote">

Method Name:
```

```
<input type="text" name="symbol" value="C">
<BR>
<input type="submit">

</FORM>
</HTML>
```

Figure 13.6 shows this HTML form in *Internet Explorer*.

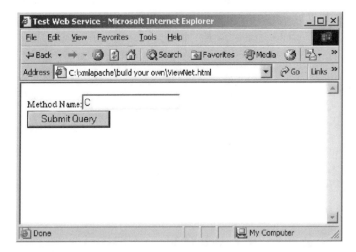

FIGURE 13.6 The HTML form that communicates with the
.NET Web Service.

.NET Web Services detect that you are communicating from HTML because
you're not using a *SOAP* request. The *.NET* Web Services then create a much simpler response, as the following code illustrates.

```
<?xml version="1.0" encoding="utf-8" ?>
<double xmlns="http://www.advocatemedia.com/">55.95</double>
```

The previous code is a simple XML document that returns with simple elements describing the type of value. Note that in the *.NET* example the method is
part of the URL rather than one of the inputs.

Using an Application to Communicate with Web Services

Java and C# contain built-in classes that make requests to Web pages and return results. These are useful for creating programs that monitor Web sites or act as your own personal spider. Almost all languages possess classes or libraries that allow you to write applications that communicate in this manner. Thus, you can use these implementations to communicate with a Web Service in a nonstandard but completely functional manner. The trick is to make the request via Get or Post from the application and then parse the response with whatever tools you have available (many include an XML parser at best and string-parsing methods at worst). Consider the following Java example.

```java
import java.net.*;
import java.io.*;

public class UseGet {

    private URL myUrl;
    private URLConnection myUrlConnection;

    //set URL
    public void setUrl(String sentUrl) {
    try {
      this.myUrl = new URL(sentUrl);
    } catch  (MalformedURLException e) {
      System.err.println(sentUrl +
      " is not a useable URL");
      System.err.println(e);
    }

  }
    //begin the connection
    public void createConnection() {
     try {
       this.myUrlConnection = myUrl.openConnection();
     } catch (Exception e) {
       System.err.println(e);
     }
    }

    //Get content of the URL, return as string
```

```
public String getUrlContent() {
  String currentLine = null;
  String returnResponse = "";

  try {
   DataInputStream webResponse =
   New DataInputStream
   (myUrlConnection.getInputStream());

   while (( currentLine = webResponse.readLine())
       != null) {
     returnResponse += currentLine;
     returnResponse += "\n";
   }
  } catch(Exception e) {
   System.err.println(e);
  }
   return(returnResponse);
}


public static void main(String args[]) {
  UseGet myUseGet = new UseGet();

  //set URL
  String someURL =
  "http://localhost:8080/axis/MoneyExchange.jws
  ?method=returnCurrencyName&countryName=Swiss";

  myUseGet.setUrl(someURL);
  myUseGet.createConnection();
  String UrlContent = myUseGet.getUrlContent();
  System.out.println(UrlContent);
}
```

In the previous Java code, there is a request made to an *Axis* Web Service via the
Get method. You know it's a Get method because in the main of the program, the
string someURL gets sent to the URL of the service and contains parameters in
the query string to specify the method and the value of parameter to pass to the
method. In this particular example, the example specifies returnCurrencyName as
the method and the parameter countryName as the value to pass to the method.

Once the URL of the service is set with the `setUrl` method, the `createConnection` method sets up the connection to the desired URL. The `getURLContent` method grabs the content and puts it into the `String UrlContent`. Then the program prints out the following XML.

```
<p>Got response message</p>
<?xml version="1.0" encoding="UTF-8"?>
<soapenv:Envelope
  xmlns:soapenv=
  "http://schemas.xmlsoap.org/soap/envelope/"
  xmls:xsd="http://www.w3.org/2001/XMLSchema"
  xmlns:xsi="http://www.w3.org/2001/XMLShema-instance">
<soapenv:Body>
  <returnCurrencyNameResponse
      soapenv:encodingStyle=
      "http://schemas.xmlsoap.org/soap/encoding/">

  <returnCurrencyNameReturn xsi:type="xsd:string">
      Franc
  </returnCurrencyNameReturn>
  </returnCurrencyNameResponse>
</soapenv:Body>
</soapenv:Envelope>
```

We can also communicate this way with *.NET* Web Services using the following URL in the same Java program.

http://www.advocatemedia.com/webservices/service1.asmx/GetTestQuote?symbol=C

The response from the *.NET* Web Service once the program executes is then the same as when we contacted it with the HTML.

```
<?xml version="1.0" encoding="utf-8"?>
<double
 xmlns="http://www.advocatemedia.com/">55.95</double>
```

Once the contents of the request make it back to the string `UrlContent` in the program, you can print it out, write it to a file, or parse it with either string or XML parsing. It all depends on what is available in your environment that doesn't support a Web Services implementation.

CONCLUSION

Although implementing a Web Service yourself appears to be a tempting process if you are stuck with a platform that is very out of date or uncommon, you'll find that you can almost always find a supported Web Services implementation that will be cheaper for you to purchase than to implement yourself. This is true because it will take thousands of developer hours to follow the appropriate standards. There are currently implementations of Web Services that support *Fortran*, *Pascal*, or even COBOL, and this probably means you can find an implementation that supports whatever strange platform you come across.

The alternative to adhering to the standards is communicating with the Web Service in a way that ignores the standards. This only works if you write software for an internal project, because external customers expect you to fully comply with all the standards. However, communicating with a Web Service in a nonstandard way may be an easy way for you to access the services you need from an unsupported platform.

A   About the CD-ROM

The CD-ROM included with *Cross-Platform Web Services Using C# and Java* includes all the code and projects from the various examples found in the book.

CD-ROM FOLDERS

Code: Contains all the code from examples in the book by chapter.

Images: Contains all the images in the book, in color, by chapter.

OVERALL SYSTEM REQUIREMENTS

- *Windows NT, Windows 2000*, or *Windows XP Pro*
  - Note that for you to use the .*NET* examples, you need one of these three operating systems so you can utilize Web services or compile code with Microsoft .*NET*. You will also need *IIS* or *Personal Web Server* installed.
  - Note that the professional versions of these operating systems will not allow you to use SSL with *IIS*, but you can use Apache with SSL on any of them.
  - For the Java examples, you can use *Windows 98, ME*, or *XP* but you will not be able to utilize any of the .*NET* examples.
- Pentium II Processor or greater
- CD-ROM drive
- Hard drive

■ 128 MBs of RAM, minimum 256 recommended
   Running both the Java Web Service examples and .*NET* examples simultaneously uses a lot of memory, so 256 MBs will make the example run faster.

■ *.NET Framework SDK/Microsoft Visual Studio.NET* recommended
   Compiling the examples in the book with the *Framework SDK* is possible, but using *Visual Studio.NET* will allow you to easily create the *Window* application examples found in the book.

■ *Java 2 SDK*
   The *SDK* contains all the components necessary to compile the examples found in the book. No visual programming environment, such as *Sun One Studio*, is required.

■ 20 MBs of hard drive space for the code examples.
   If you download much of the software mentioned in Appendix B you may need up to 300 MBs to handle *Visual Studio.NET* and other large installations.

B Software Used in This Book

This book utilizes a lot of different software to make cross-platform Web Services work. This appendix focuses on where to download the various software needed to use the examples found on the CD-ROM.

MICROSOFT PRODUCTS

Unfortunately, there is no free download for *Visual Studio.NET*, but Microsoft does provide the *.NET Framework SDK*, which includes all the needed functionality to compile the examples on the CD-ROM. *Visual Studio* offers the advantage that it makes things like creating a Web reference easier or making a *Windows* application quicker. Thus, it is strongly recommended that you purchase *Visual Studio.NET* to utilize the examples found in the book.

.NET Framework SDK

Available at:
http://msdn.microsoft.com/downloads/default.asp?url=/downloads/topic.asp?URL=/ MSDN-FILES/028/000/123/topic.xml
Contains access to all the command line versions of the compiler for all the languages involved in *.NET* including C# and *Visual Basic.NET*. It also has tools for creating proxies and WSDL for Web Services.

.NET Framework Redistributable

Also available at the previous URL.
Only contains enough of the *.NET* framework to execute code created within the *SDK* or *Visual Studio.NET*. Although not required for any of the examples in

the book, it may be useful for you to deploy any Web Services you create to a production server.

.NET WebService Studio

Available at:

http://www.gotdotnet.com/team/tools/web_svc/default.aspx

This is a tool that is an open source piece of software available at *www.gotdotnet.com*. Even though it comes from this Web site, all the contact information related to it is to Microsoft. So it is essentially a Microsoft product.

This is a great tool because it compiles the WSDL, creates a proxy, and then allows you to test client code against the proxy. This saves time because you don't need to write a consumer from scratch to test your Web Service. It is also useful to test any Web Service that is Java based and that produces WSDL.

Web Services Development Toolkit

Available at:

http://msdn.microsoft.com/downloads/default.asp?URL=/downloads/sample.asp?url= /MSDN-FILES/027/001/997/msdncompositedoc.xml

This is an early software release from Microsoft that utilizes the up-and-coming security methods for Web Services. This includes using an X509 certificate to sign a Web Services request.

UDDI SDK

Available at:

http://msdn.microsoft.com/downloads/default.asp?URL=/downloads/sample.asp?url= /msdn-files/027/001/814/msdncompositedoc.xml

Provides an interface to the *UDDI* standard so that you can write applications and dynamic Web pages that query the repository.

APACHE DOWNLOADS

The Apache Web Services tools require multiple downloads to get them to work. The price for using shareware is that you don't get a simple installation program.

The installation of both Apache *SOAP* and Apache *Access* is not difficult but it is time consuming.

Follow the directions for installation found in Chapters 7 and 8.

Apache *Axis*

Available at:

http://xml.apache.org/axis/index.html

This is a complete rewrite of the Apache *SOAP* library. Many of the features make it easier to integrate with *.NET*. Many of the features also provide functionality to make it much easier to call Web Services in general.

Apache *SOAP*

Available at:

http://xml.apache.org/soap/index.html

Although this edition is out of date, Apache *Axis*, at the time of this writing, still isn't officially released. Therefore, you may be required to use this version until the first release of *Axis*. You may also encounter many legacy applications written with Apache *SOAP*.

Apache *Tomcat*

Available at:

http://jakarta.apache.org/site/binindex.html

Tomcat runs all the server-side Java code that appears in this book. The author used *Tomcat* Version 3.3 because that appears to be the most stable, but any version greater than Version 3 should work.

Apache *Xerces*

Available at:

http://xml.apache.org/dist/xerces-j/

This is the XML parser that allows both *Axis* and *SOAP* to parse the XML in *SOAP* requests and responses along with the WSDL. Remember that you need Version 1.4.4 rather than one of the newer versions, such as Version 2.0.

Apache *Web Server with SSL Encryption*

Available at:

http://www.openssl.org

This version of Apache allows you to utilize SSL and create Certificate Signing Requests, which is important to the security examples in Chapter 10.

SUN DOWNLOADS

Sun Microsystems provides several tools that make the Java Web Services shown in this book possible. The most important of these downloads is Java, but the Java *Activation Framework* and *JavaMail* are also important.

Java Version 1.4.0

Available at:

http://java.sun.com/j2se/

The author used Version 1.4.0 to compile the examples found in the book.

JavaBeans *Activation Framework*

Available at:

http://java.sun.com/products/javabeans/glasgow/jaf.html

The JavaBeans *Activation Framework* is required by both Apache *SOAP* and *Axis*.

JavaMail

Available at:

http://java.sun.com/products/javamail/

An API that allows you to send attachments and manipulate e-mail through Java. *Axis* uses this to send attachments through *SOAP* messages.

C Apache License

Many of the examples and tools used throughout this book were inspired by several examples found within much of the Apache software packages. The license is included for you to have a complete understanding of how open Apache is to consumers using their products.

```
The Apache Software License, Version 1.1


Copyright (c) 1999-2000 The Apache Software Foundation. All rights
reserved.

Redistribution and use in source and binary forms, with or without
modification, are permitted provided that the following conditions
are met:

1. Redistributions of source code must retain the above copyright
   notice, this list of conditions and the following disclaimer.

2. Redistributions in binary form must reproduce the above copyright
   notice, this list of conditions and the following disclaimer in
   the documentation and/or other materials provided with the
   distribution.

3. The end-user documentation included with the redistribution,
   if any, must include the following acknowledgment:
   "This product includes software developed by the
   Apache Software Foundation (http://www.apache.org/)."
   Alternately, this acknowledgment may appear in the software itself,
   if and wherever such third-party acknowledgments normally appear.

4. The names "Xerces" and "Apache Software Foundation" must
   not be used to endorse or promote products derived from this
```

software without prior written permission. For written permission, please contact apache@apache.org.

5. Products derived from this software may not be called "Apache," nor may "Apache" appear in their name, without prior written permission of the Apache Software Foundation.

THIS SOFTWARE IS PROVIDED ``AS IS'' AND ANY EXPRESSED OR IMPLIED WARRANTIES, INCLUDING, BUT NOT LIMITED TO, THE IMPLIED WARRANTIES OF MERCHANTABILITY AND FITNESS FOR A PARTICULAR PURPOSE ARE DISCLAIMED. IN NO EVENT SHALL THE APACHE SOFTWARE FOUNDATION OR ITS CONTRIBUTORS BE LIABLE FOR ANY DIRECT, INDIRECT, INCIDENTAL, SPECIAL, EXEMPLARY, OR CONSEQUENTIAL DAMAGES (INCLUDING, BUT NOT LIMITED TO, PROCUREMENT OF SUBSTITUTE GOODS OR SERVICES; LOSS OF USE, DATA, OR PROFITS; OR BUSINESS INTERRUPTION) HOWEVER CAUSED AND ON ANY THEORY OF LIABILITY, WHETHER IN CONTRACT, STRICT LIABILITY, OR TORT (INCLUDING NEGLIGENCE OR OTHERWISE) ARISING IN ANY WAY OUT OF THE USE OF THIS SOFTWARE, EVEN IF ADVISED OF THE POSSIBILITY OF SUCH DAMAGE.
==

This software consists of voluntary contributions made by many individuals on behalf of the Apache Software Foundation and was originally based on software copyright (c) 1999, International Business Machines, Inc., http://www.ibm.com. For more information on the Apache Software Foundation, please see <http://www.apache.org/>.

D

Visual Basic.NET

Visual Basic.NET (*VB.NET*) is also a part of Microsoft's *.NET* initiative that supports Web Services. There are millions of developers who use *VB 6*, the predecessor to *VB.NET*, every day who are considering upgrading to *.NET*. This appendix examines creating Web Services and consumers using *Visual Basic* coding. Note that the download for the *.NET Framework SDK* mentioned in Appendix B does include a command line version of the *VB.NET* compiler.

CREATING A *VB.NET* WEB SERVICE

Open *Visual Studio.NET* and create a new project. From the "New Project" dialogue, select "Visual Basic Projects" on the lefthand side and then on the right select "*ASP.NET* Web Service." Figure D.1 shows the appropriate selections.

The code *Visual Studio.NET* generates for you looks like the following.

```
Imports System.Web.Services

<WebService(Namespace := "http://tempuri.org/")> _
Public Class XPlatform
Inherits System.Web.Services.WebService

#Region " Web Services Designer Generated Code "

Public Sub New()
    MyBase.New()

    'This call is required by the Web Services
    'Designer.
    InitializeComponent()
```

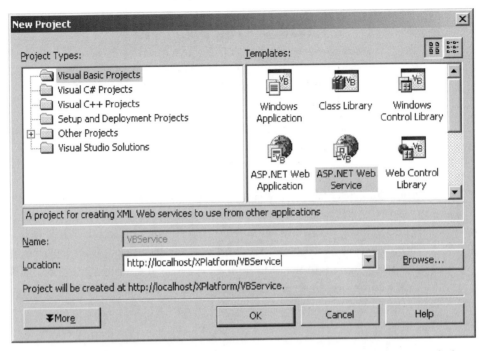

FIGURE D.1 The "New Project" dialogue in *Visual Studio.NET* with selections made for *Visual Basic.NET*.

```vb
        'Add your own initialization code after the
        'InitializeComponent() call

    End Sub

    'Required by the Web Services Designer
    Private components As System.ComponentModel.IContainer

        'NOTE: The following procedure is required by the
        'Web Services
        'Designer
        'It can be modified using the Web Services
        'Designer.
        'Do not modify it using the code editor.
        <System.Diagnostics.DebuggerStepThrough()>
        Private Sub InitializeComponent()
        components = New System.ComponentModel.Container()
```

```
End Sub

Protected Overloads Overrides Sub Dispose
    (ByVal disposing As Boolean)
    'CODEGEN: This procedure is required by the Web
    'Services Designer
    'Do not modify it using the code editor.
    If disposing Then
        If Not (components Is Nothing) Then
            components.Dispose()
        End If
    End If
    MyBase.Dispose(disposing)
End Sub

#End Region

' WEB SERVICE EXAMPLE
' The HelloWorld() example service returns the string
' Hello World.
' To build, uncomment the following lines then save and
' build the project.
' To test this web service, ensure that the .asmx file
' is the start page
' and press F5.
'
'<WebMethod()> Public Function HelloWorld() As String
'   HelloWorld = "Hello World"
' End Function

End Class
```

In the previous example, all the code that occurs before the `#End Region` is generated by *Visual Studio.NET* when you create the Web Service project. Much of this code calls many of the same objects as the code that gets generated when you create a Web Service in C#. Refer to Chapter 6 to see this code generated in a C# Web Service.

To add a method, remove the commented out `hello world` *VS.NET* example and add the following code.

```
<WebMethod(Description:="Simple Stock Example in " _
   "VB.NET")> _
```

```
Public Function GetTestQuote(ByVal symbol As String)_
As Double
    Dim stockValue As Double = 55.95
    Dim empty As Double = -1

    If symbol = "C" Then
        Return (stockValue)
    Else
        Return (empty)
    End If

End Function
```

C# programmers resist the urge to use semicolons and open and closing brackets ({}) in the Visual Basic *code.* Visual Basic *doesn't have its roots in C and C++ like C#, so the syntax isn't similar. Notice how the return type is last in the* Function *definition and how the if statement ends with an* End If. *This type of syntax has its roots in the* Basic *programming language, which the author programmed in as a kid on his Apple IIe.*

If you ignore the syntax of the Web Service, you'll see some similarities. You still needed to define the function as a WebMethod and provide a signature for the function.

CREATING A *VB.NET* CLIENT

To create a *VB.NET* consumer you need to create the proxy code. Just as in C#, you can either add a Web reference through *VS.NET* or use the WSDL.exe tool to create the proxy. The commands are exactly the same as for C#, but with the *WSDL.exe* tool, you need to specify *Visual Basic* as the language you want the proxy in.

Even if the proxy gets created with C# code and then compiled with the C# compiler, you can still include and use the proxy with VB.NET. This is one of the benefits of the Common Language Runtime.

Like C#, *Visual Basic.NET* has the ability to create *ASP.NET* pages, command line applications, and *Windows* GUI tools—all of which can create a proxy and then access the `SimpleStockQuote` example with the following code.

```
Dim localhostWebservice As Xplatform
Double result = localhostWebservice.GetTestQuote("C")
```

Review Chapter 6 to get an idea of the types of consumers *Visual Basic.NET* can support. Any consumer you can create in C# can also be created in *VB.NET*.

E Using PERL to Access Web Services

PERL is a scripting language that is similar to C. It was one of the frontier Web languages that allowed interactive Web pages through the Common Gateway Interface (CGI). PERL's role as a Web language has started to fade with the advent of *J2EE*, *.NET*, *Cold Fusion*, and other technologies, but PERL is still a superior language for automating tasks and performing system administration duties.

PERL's main advantage is that it is a scripting language that makes it easy to change, compile, and execute. In fact, it's a great language to use when performing software testing because of its ease of use and its ability to easily manipulate and parse text files. This gives you the ability to easily search for results in a generated log file.

To call Web Services from PERL, all you need to do is write a PERL script that uses the SOAP::Lite Library available from *www.soaplite.com*. The following code example contains a call to the returnCurrencyName method from the MoneyExchange Web Service covered in Chapter 11. The script begins by using a bracket to begin the scope of the program. Then it calls the SOAP::Lite Library with the use statement. Then a print method is associated with the SOAP::Lite call so the output from the method is displayed at the command line when the script executes. The location of the WSDL file of the service you wish to call gets passed to the service attribute. Then, the country whose currency name you are querying gets passed to the method from the service, returnCurrencyName. Notice that we didn't need to create a proxy for PERL because it parses the WSDL on the fly to learn how to communicate with the service.

```
{#begin scope PERL program

use SOAP::Lite;

#Call Axis Service
```

ope.

```
    print "Name of currency in Switzerland is: ";
    print SOAP::Lite
    -> service
    ('http://localhost:8080/axis/MoneyExchange.jws?wsdl')
    -> returnCurrencyName('Swiss');

}#end scope PERL Program
```

The next PERL example calls three of the example Web Services found in various chapters of this book to show that PERL can call Web Services from *SOAP*, *Axis*, and *.NET*.

```
{#begin scope PERL program

use SOAP::Lite;

 #Call Axis Service
 print "Name of currency in Switzerland is: ";
 print SOAP::Lite
  -> service
 ('http://localhost:8080/axis/MoneyExchange.jws?wsdl')
  -> returnCurrencyName('Swiss');

 #add return line
 print "\n";


 #Call Apache SOAP Service
 print "Type of Service being called: ";
 print SOAP::Lite
 -> service('http://localhost:8080/soap/GetId.wsdl')
 -> ServiceId();

#Add Return Line
print "\n";

#Call .NET Web Service
print "The Value of C in simple stock quote is ";
print SOAP::Lite
-> service(
 'http://localhost/XPlatform/StockQuote.asmx?wsdl'
 )
```

```
-> GetTestQuote("C");

}#end scope PERL program
```

As you can see, PERL calls Web Services fairly easily, and this provides one more tool for testing and communicating with Web Services.

F Microsoft's *UDDI .NET SDK*

Microsoft provides a Beta version of an API that allows you to interact with a UDDI Web site using an application or Web page that you create. The example provided in this appendix illustrates how to use the API to interact with Microsoft's UDDI Web site. The C# code is put into a *Windows* GUI, but it easily could have been put inside an *ASP.NET* page or a command line application.

First, download the *SDK* from the location mentioned in Appendix B, and install it according to the instructions provided on Microsoft's Web site. Create a new C# *Windows* application, and to utilize the functionality in the *SDK* you need to add a reference to the `Microsoft.Uddi.Sdk.dll` that came with the download. This is done by creating a new project in *Visual Studio.NET* and going to the "Project" menu and selecting "Add Reference." Then use the "Browse" option to select the dll. Figure F.1 shows the dialogue.

Create a *Windows* application that has a button, a text box, and a couple of labels. Figure F.2 shows how you might want the GUI to appear for the UDDI. Note that in this image the GUI has already made the call to the UDDI Web site.

Most of the important code that occurs in this example is found within the "Click Event" of the button in the upper right corner, and this is the following example. The first step is to tell the code where the inquire URL resides for this particular UDDI implementation. In the case of Microsoft's UDDI Web site, that URL is *http://uddi.microsoft.com/inquire*.

The next step involves creating a `FindBusiness` object. This is the object that actually makes the request and receives the data. The next step sets the `Name` attribute for the `FindBusiness` object so the request knows what information to ask the UDDI site for. Then a `BusinessList` object is created and this is a result set for the UDDI request.

To get all the results, you need to move through the `BusinessList` object `myBusiness` as if it were a result set from a database. The first piece of data to request

369

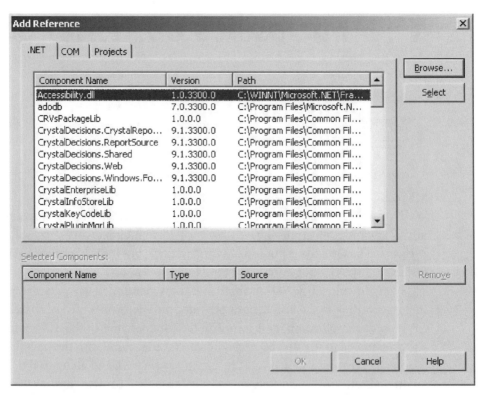

FIGURE F.1 Adding a reference to a project within *Visual Studio.NET*.

from the object is the Name and Descriptions of the company we ask for information about. Then there is a loop that displays all the keys and descriptions for the
entries of each company. These values are all appended to the String Results, and
once the loops are complete all the information in Results gets sent to the label
uddiDisplay.

```
private void button1_Click(object sender, System.EventArgs e)
{
  try
    {
    //Call the UDDI repository
    Inquire.Url = "http://uddi.microsoft.com/inquire";

    //Create a find business object
```

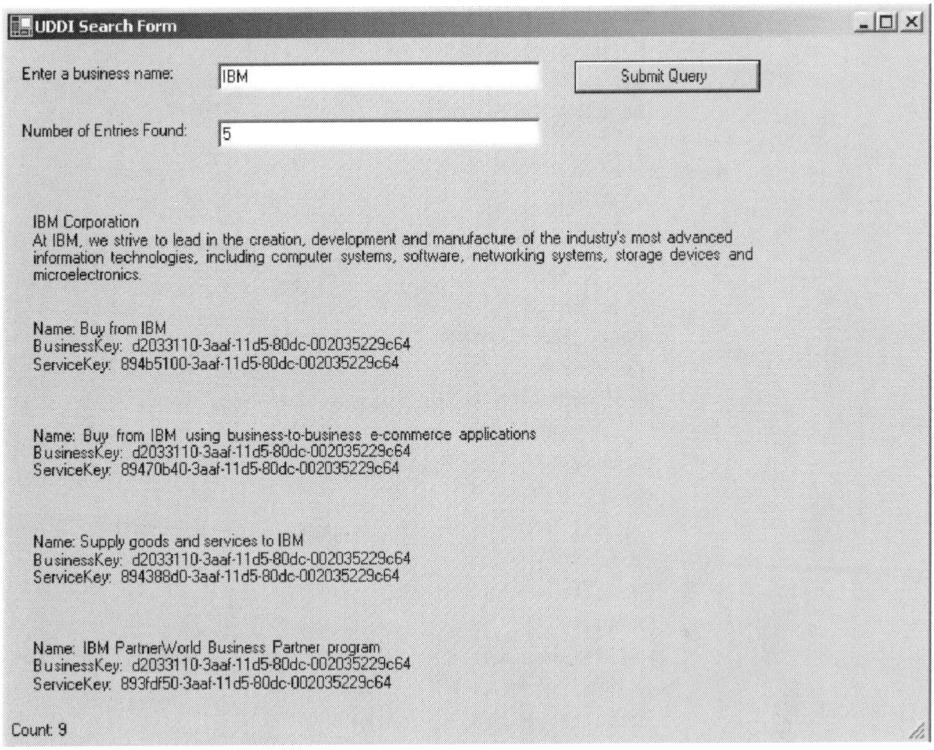

FIGURE F.2 How the GUI for the *UDDI SDK* example in this appendix may appear.

```
FindBusiness myFindBusiness = new FindBusiness();
myFindBusiness.Name = businessEntry.Text;

BusinessList myBusinessList = myFindBusiness.Send();
entryCount.Text =
myBusinessList.BusinessInfos.Count.ToString();

if (myBusinessList.Truncated ==
 Microsoft.Uddi.Api.TruncatedEnum.True)
   ErrorBar.Text = "Results are truncated";

   String Results = null;

   if (myBusinessList.BusinessInfos.Count > 0)
   {
```

```
          Results += myBusinessList.BusinessInfos[0].Name;
          Results += "\n" +
          myBusinessList.BusinessInfos[0].Descriptions[0].Text+"\n";
          Results += "\n\n";
          }

          for (int j=0;
              j<myBusinessList.BusinessInfos[0].ServiceInfos.Count;
              j++)
          {
          Results += "Name: ";
          Results+=
              myBusinessList.BusinessInfos[0].ServiceInfos[j].Name +
              "\n";
          Results += "BusinessKey: ";
          Results +=
          myBusinessList.BusinessInfos[0].ServiceInfos[j].BusinessKey
          + "\n";
          Results += "ServiceKey: ";
          Results +=
          myBusinessList.BusinessInfos[0].ServiceInfos[j].ServiceKey
          + "\n";
          Results += "\n\n\n";
          ErrorBar.Text = "Count: " + j;
          }
          //display value of Results
          uddiDisplay.Text = Results;
          }
     catch (UddiException error)
     {
     ErrorBar.Text = " " + error.Number + " "
     + error.Message;
     }
     catch (Exception error)
     {
     ErrorBar.Text = error.Message;
     }


     }
```

The beauty of this *SDK* is that you could write an application that searches for an appropriate Web Service and then automatically creates the proxy and you then have access to its method. This code also allows you to create Web pages internally for your corporation that can search the UDDI repository without your users having to figure out which one your corporation supports.

Index